# WORKSHEETS
## FOR CLASSROOM OR LAB PRACTICE

*with contributions from*

BEVERLY FUSFIELD

STEVE OUELLETTE

JAMES J. BALL
*Indiana State University*

# DEVELOPMENTAL MATHEMATICS:
# BASIC MATHEMATICS
# AND ALGEBRA

## SECOND EDITION

### Margaret L. Lial
*American River College*

### John Hornsby
*University of New Orleans*

### Terry McGinnis

### Stanley A. Salzman
*American River College*

### Diana L. Hestwood
*Minneapolis Community and Technical College*

**Addison-Wesley**
is an imprint of

Copyright © 2010 Pearson Education, Inc.
Publishing as Pearson Addison-Wesley, 75 Arlington Street, Boston, MA 02116.

ISBN-13: 978-0-321-59973-5
ISBN-10: 0-321-59973-X

1 2 3 4 5 6 OPM 12 11 10 09 08

**Addison-Wesley**
is an imprint of

PEARSON

www.pearsonhighered.com

# CONTENTS

# Chapter 1 WHOLE NUMBERS

### 1.1    Reading and Writing Whole Numbers

| Learning Objectives |
| --- |
| 1    Identify whole numbers. |
| 2    Give the place value of a digit. |
| 3    Write a number in words or digits. |
| 4    Read a table. |

## Key Terms

Use the vocabulary terms listed below to complete each statement in exercises 1–3.

**whole numbers**          **place value**          **table**

1.   A display of facts in rows and columns is called a _____.

2.   The _____ are 0, 1, 2, 3, 4, and so on.

3.   The _____ of each digit in a whole number is determined by its position in the whole number.

### Objective 1    Identify whole numbers.

*Indicate whether each number is a whole number or not a whole number.*

1.   48                                              1. _____

2.   1.2                                             2. _____

3.   $6\frac{3}{4}$                                 3. _____

4.   10,029                                          4. _____

### Objective 2    Give the place value of a digit.

*Write the digit for the given place value in each of the following whole numbers.*

5.   9841          thousands                          5. _____

                   tens                                 _____

6.   25,016        ten-thousands                      6. _____

                   hundreds                             _____

7.   186,321       hundred-thousands                  7. _____

                   ones                                 _____

**8.** 5,813,207        millions                    8. _____

                       thousands                    _____

**9.** 2,800,439,012    billions                     9. _____

                       ten-millions                 _____

*Write the digits for the given period (group) in each whole number*

**10.** 29,176          thousands                   10. _____

                        ones                        _____

**11.** 75,229,301      millions                    11. _____

                        thousands                   _____

                        ones                        _____

**12.** 70,000,603,214  billions                    12. _____

                        millions                    _____

                        thousands                   _____

                        ones                        _____

**13.** 300,459,200,005 billions                    13. _____

                        millions                    _____

                        thousands                   _____

                        ones                        _____

## Objective 3    Write a number in words or digits

*Write each number in words.*

**14.** 8714                                         14. _____

**15.** 39,015                                       15. _____

**16.** 834,768                                      16. _____

**17.** 2,015,102                    **17.** _____

**18.** 499,304,018                  **18.** _____

*Write each number using digits.*

**19.** Four thousand, one hundred twenty-seven        **19.** _____

**20.** Twenty-nine thousand, five hundred sixteen     **20.** _____

**21.** Six hundred eight-five million, two hundred fifty-nine        **21.** _____

**22.** Three hundred million, seventy-five thousand, two        **22.** _____

*Write the numbers from each sentence using digits.*

**23.** A bottle of a certain vaccine will give seven thousand, two hundred ten injections.        **23.** _____

**24.** Every year, nine hundred seventy-two thousand, four hundred thirty people visit a certain historical area.        **24.** _____

**25.** A supermarket has fifteen thousand three hundred thirteen different items for sale.        **25.** _____

**26.** The population of a large city is six million, two hundred five thousand.        **26.** _____

Name: _____  Date: _____

Instructor: _____  Section: _____

## Objective 4   Read a table.

*Use the table below for exercises 27–30. Write the number in digits.*

| Weight of Exerciser | 123 lbs | 130 lbs | 143 lbs |
|---|---|---|---|
| Calories burned in 30 minutes | | | |
| Cycling | 168 | 177 | 195 |
| Running | 324 | 342 | 375 |
| Jumping Rope | 273 | 288 | 315 |
| Walking | 162 | 171 | 189 |

**Source: Fitness magazine**

27. The number of calories burned by a 130-pound adult in 30 minutes of cycling.

27. _____

28. The activity which will burn at least 300 calories when performed by a 123-pound adult.

28. _____

29. The number of calories burned by a 143-pound adult in 30 minutes of jumping rope.

29. _____

30. The weight of an adult who will burn 189 calories in 30 minutes of walking.

30. _____

4

# Chapter 1 WHOLE NUMBERS

### 1.2    Adding Whole Numbers

| **Learning Objectives** |
| :--- |
| 1    Add two single-digit numbers. |
| 2    Add more than two numbers. |
| 3    Add when regrouping (carrying) is not required. |
| 4    Add with regrouping (carrying). |
| 5    Use addition to solve application problems. |
| 6    Check the answer in addition. |

### Key Terms

Use the vocabulary terms listed below to complete each statement in exercises 1–7.

**addition      addends      sum (total)      commutative property of addition**

**associative property of addition          regrouping          perimeter**

1.   By the_____, changing the order of the addends in an addition problem does not change the sum.

2.   In addition, the numbers being added are called the _____.

3.   The process of finding the total is called _____.

4.   By the_____, changing the grouping of the addends in an addition problem does not change the sum.

5.   If the sum of the digits in any column is greater than 9, use the process called _____.

6.   The distance around the outside edges of a figure is called the _____.

7.   The answer to an addition problem is called the _____.

### Objective 1    Add two single-digit numbers.

*Add.*

1.   $3 + 9$                                    1. _____

2.   $8 + 7$                                    2. _____

**Objective 2    Add more than two numbers.**

*Add.*

3.      7
        2
        5
        3
      + 8

3. _____

4.      9
        7
        2
        5
      + 6

4. _____

5.      3
        4
        9
        2
        7
      + 6

5. _____

6.      6
        4
        9
        8
        4
      + 3

6. _____

**Objective 3    Add when regrouping (carrying) is not required.**

*Add.*

7.     42
     + 57

7. _____

**8.**      421
        + 567

**8.** _____

**9.**      86,305
        + 12,672

**9.** _____

**10.**     45,158
            20,340
        +    2401

**10.** _____

## Objective 4    Add with regrouping (carrying).

*Add.*

**11.**      83
         + 29

**11.** _____

**12.**      563
         + 478

**12.** _____

**13.**      7439
         + 8376

**13.** _____

**14.**      7033
             809
             2532
         +    41

**14.** _____

**15.**    3197 + 420 + 638 + 67

**15.** _____

**16.**    6835 + 97 + 246 + 4001

**16.** _____

**Objective 5    Use addition to solve application problems.**

*Using the map below, find the shortest distance between the following cities. All distances are in miles.*

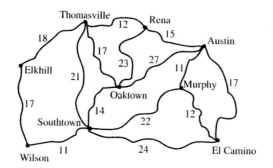

**17.**    Murphy and Thomasville

17. _____

**18.**    Wilson and Austin

18. _____

**19.**    El Camino and Thomasville

19. _____

*Solve the following application problems, using addition.*

**20.**    Kevin Levy has 52 nickels, 37 dimes, and 119 quarters. How many coins does he have altogether?

20. _____

**21.**    The theater sold 276 adult tickets, and 349 child tickets. How many tickets were sold altogether?

21. _____

**22.**    At a charity bazaar, a church has a total of 1873 books for sale, while a lodge has 3358 books for sale. How many books are for sale?

22. _____

Name:                   Date:

Instructor:           Section:

*Find the perimeter or total distance around each of the following figures.*

**23.**

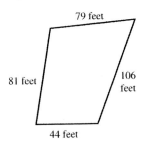

**23.** _____

**24.**

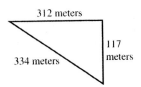

**24.** _____

**25.**

206 yards   197 yards

107 yards              107 yards

427 yards

**25.** _____

## Objective 6     Check the answer in addition.

*Check the following additions. If an answer is incorrect, give the correct answer.*

**26.**
$$\begin{array}{r} 67 \\ 48 \\ +\ 83 \\ \hline 198 \end{array}$$

**26.** _____

**27.**
$$\begin{array}{r} 73 \\ 9815 \\ 390 \\ +\ 7002 \\ \hline 16,270 \end{array}$$

**27.** _____

**28.**
```
    723
    681
     29
    412
 + 103
 ─────
   1947
```

**28.** _____

**29.**
```
   3028
    335
   2914
    688
 + 1647
 ──────
   8612
```

**29.** _____

**30.**
```
     72
     38
   5735
    764
 +   16
 ──────
   6625
```

**30.** _____

# Chapter 1 Whole Numbers

### 1.3    Subtracting Whole Numbers

| **Learning Objectives** | |
| --- | --- |
| 1 | Change addition problems to subtraction and subtraction problems to addition. |
| 2 | Identify the minuend, subtrahend, and difference. |
| 3 | Subtract when no regrouping (borrowing) is needed. |
| 4 | Check subtraction answers by adding. |
| 5 | Subtract by regrouping (borrowing). |
| 6 | Solve application problems with subtraction. |

### Key Terms

Use the vocabulary terms listed below to complete each statement in exercises 1–4.

**minuend      subtrahend         difference    regrouping**

1.  The number from which another number is being subtracted is called the

    _____ .

2.  In order to subtract 29 from 76, use a process called _____ from
    the tens place.

3.  The _____ is the number being subtracted.

4.  The answer to a subtraction problem is called the _____ .

### Objective 1    Change addition problems to subtraction and subtraction problems to addition.

*Write two subtraction problems for each addition problem.*

1.  $149 + 38 = 187$                                  1. _____

2.  $478 + 239 = 717$                                 2. _____

*Write an addition problem for each subtraction problem.*

3.  $1211 - 426 = 785$                                3. _____

4.  $5094 - 113 = 4981$                               4. _____

## Objective 2    Identify the minuend, subtrahend, and difference.

*Identify the minuend, subtrahend, and difference in each of the following subtraction problems.*

5.    $98 - 36 = 62$

**5.**

Minuend _____

Subtrahend _____

Difference _____

6.    $35 - 9 = 24$

**6.**

Minuend _____

Subtrahend _____

Difference _____

## Objective 3    Subtract when no regrouping (borrowing) is needed.

*Subtract.*

7.    $\begin{array}{r} 5573 \\ -\ 422 \\ \hline \end{array}$

**7.** _____

8.    $\begin{array}{r} 8539 \\ -\ 2527 \\ \hline \end{array}$

**8.** _____

## Objective 4    Check subtraction answers by adding.

*Check the following subtractions. If an answer is not correct, give the correct answer.*

9.    $\begin{array}{r} 192 \\ -\ 39 \\ \hline 167 \end{array}$

**9.** _____

10.    $\begin{array}{r} 4847 \\ -\ 3768 \\ \hline 1121 \end{array}$

**10.** _____

**11.**     5763
         – 2783
         3980

**11.** _____

**12.**     31,146
         – 7312
         23,834

**12.** _____

**13.**     82,004
         – 3917
         79,193

**13.** _____

**Objective 5    Subtract by regrouping (borrowing).**

_Subtract._

**14.**     927
         – 729

**14.** _____

**15.**     613
         – 421

**15.** _____

**16.**     4687
         – 2769

**16.** _____

**17.**     33,728
         – 7829

**17.** _____

**18.**     86,372
         – 29,485

**18.** _____

**19.**     302
         – 57

**19.** _____

**20.**    7000
          − 297                            **20.** _____


## Objective 6    Solve application problems with subtraction.

*Solve each application problem.*

**21.**    A Girl Scout has 52 boxes of cookies to sell. If she    **21.** _____
          sells 27 boxes, how many boxes will she have left?


**22.**    An airplane is carrying 234 passengers. When it        **22.** _____
          lands in Atlanta, 139 passengers get off the plane.
          How many passengers are then left on the plane?


**23.**    Nathaniel Best has $553 in his checking account. He    **23.** _____
          writes a check for $134. How much is then left in the
          account?


**24.**    On Sunday, 7342 people went to a football game,        **24.** _____
          while on Monday, 9138 people went. How many
          more people went on Monday?


**25.**    Sally Tanner had $22,143 withheld from her             **25.** _____
          paycheck last year for income tax. She actually owes
          only $16,959 in tax. What refund should she
          receive?


**26.**    The Conrads now pay $439 per month for rent. If       **26.** _____
          they rent a larger apartment, the payment will be
          $702 per month. How much extra will they pay each
          month?

**27.** On Friday, 11,594 people visited Eastridge Amusement Park, while 14,352 people visited the park on Saturday. How many more people visited the park on Saturday?

27. _____

**28.** One bid for painting a house was $2134. A second bid was $1954. How much would be saved using the second bid?

28. _____

**29.** Last year, 574 athletes competed in a district tract meet at Johnson College. This year, 498 athletes competed. How many fewer athletes competed this year than last?

29. _____

**30.** At People's Bank, Marc Lukas can earn $1538 per year in interest, while Farmer's Bank would pay him $1643 interest. How much additional interest would he earn at the second bank?

30. _____

# Chapter 1 WHOLE NUMBERS

## 1.4    Multiplying Whole Numbers

| Learning Objectives | |
|---|---|
| 1 | Identify the parts of a multiplication problem. |
| 2 | Do chain multiplications. |
| 3 | Multiply by single-digit numbers. |
| 4 | Use multiplication shortcuts for numbers ending in zeros. |
| 5 | Multiply by numbers having more than one digit. |
| 6 | Solve application problems with multiplication. |

### Key Terms

Use the vocabulary terms listed below to complete each statement in exercises 1–6.

**factors        product        commutative property of multiplication**

**associative property of multiplication        chain multiplication problem**

**multiple**

1.    By the_____, changing the order of the factors in a multiplication problem does not change the product.

2.    In multiplication, the numbers being multiplied are called the

    _____.

3.    By the_____, changing the grouping of the factors in a multiplication problem does not change the product.

4.    The product of two whole number factors is called a _____ of either factor.

5.    The answer to a multiplication problem is called the _____.

6.    A multiplication problem with more than two factors is a

    _____.

### Objective 1    Identify the parts of a multiplication problem.

*Identify the factors and the product in each multiplication problem.*

1.    $5(2) = 10$

**1.**
**Factors** _____

**Product** _____

**2.**     $108 = 9 \times 12$

**2.**

**Factors** _____

**Product** _____

**Objective 2**     **Do chain multiplications.**

*Multiply.*

  **3.**     $4 \times 4 \times 2$

                                                    **3.** _____

  **4.**     $3 \times 4 \times 7$

                                                    **4.** _____

  **5.**     $(6)(4)(8)$

                                                    **5.** _____

**Objective 3**     **Multiply by single-digit numbers.**

*Multiply.*

  **6.**      54

           $\times\ 4$

                                                    **6.** _____

  **7.**     163

          $\times\ \ 5$

                                                   **7.** _____

  **8.**     405

          $\times\ \ 7$

                                                   **8.** _____

  **9.**     31,763

         $\times\ \ \ \ \ 9$

                                                   **9.** _____

**10.**     30,009

         $\times\ \ \ \ \ 6$

                                                **10.** _____

**Objective 4    Use multiplication shortcuts for numbers ending in 0s.**

*Multiply.*

11.    $439 \times 1000$                              11. _____

12.    $(852)(30)$                                   12. _____

13.    $3005 \times 2000$                            13. _____

14.    $500 \times 40$                               14. _____

15.    $8234 \times 2000$                            15. _____

**Objective 5    Multiply by numbers having more than one digit.**

*Multiply.*

16.       $\begin{array}{r} 644 \\ \times\ 19 \\ \hline \end{array}$                16. _____

17.       $\begin{array}{r} 4031 \\ \times\ \ 48 \\ \hline \end{array}$               17. _____

**18.**      7165

          $\times$   53

**18.** _____

**19.**      5249

          $\times$   63

**19.** _____

**20.**      8621

          $\times$   131

**20.** _____

## Objective 6    Solve application problems with multiplication.

*Solve the following application problems.*

**21.**     A fabric store has 16 bolts of silk. Each bolt contains 35 yards of silk. How many yards of silk does the fabric store have in all?

**21.** _____

**22.**     On a recent trip the Jensen family drove 45 miles per hour on the average. They drove 22 hours altogether. How many miles did they drive altogether?

**22.** _____

**23.**     Marisa Taylor saves $38 out of every pay check. Last year she received 24 pay checks. How much did she save?

**23.** _____

**24.** Heinen's Supermarket received a shipment of 28 cartons of canned vegetables. There were 24 cans in each carton. How many cans were there altogether?

24. _____

**25.** The 2008 Toyota Prius is estimated to get 48 miles per gallon in city driving. Its fuel tank holds approximately 12 gallons of gasoline. How far can it travel on one tank of gasoline?

25. _____

*Find the total cost of the following items.*

**26.** 18 chairs at $42 per chair

26. _____

**27.** 512 boxes of chalk at $19 per box

27. _____

**28.** 47 watches at $29 per watch

28. _____

**29.** 178 baseball caps at $9 per cap

29. _____

**30.** 79 clocks at $198 per clock

30. _____

# Chapter 1 WHOLE NUMBERS

## 1.5 Dividing Whole Numbers

| **Learning Objectives** | |
|---|---|
| 1 | Write division problems in three ways. |
| 2 | Identify the parts of a division problem. |
| 3 | Divide 0 by a number. |
| 4 | Recognize that a number cannot be divided by 0. |
| 5 | Divide a number by itself. |
| 6 | Divide a number by 1. |
| 7 | Use short division. |
| 8 | Use multiplication to check the answer to a division problem. |
| 9 | Use tests for divisibility. |

### Key Terms

Use the vocabulary terms listed below to complete each statement in exercises 1–5.

**dividend      divisor      quotient      short division      remainder**

1.  The number left over when two numbers do not divide exactly is the

    _____.

2.  The number being divided by another number in a division problem is the

    _____.

3.  The answer to a division problem is called the _____.

4.  In the problem $639 \div 9$, 9 is called the _____.

5.  _____ is a method of dividing a number by a one-digit divisor.

### Objective 1    Write division problems in three ways.

*Write each division problem using two other symbols.*

1.    $15 \div 3 = 5$                                          1. _____

2.    $\dfrac{50}{25} = 2$                                    2. _____

## Objective 2    Identify the parts of a division problem.

*Identify the dividend, divisor, and quotient.*

3.    $63 \div 7 = 9$

**3.**
**Dividend** _____

**Divisor** _____

**Quotient** _____

4.    $5\overline{)30}$ with $6$ on top

**4.**
**Dividend** _____

**Divisor** _____

**Quotient** _____

5.    $\dfrac{44}{11} = 4$

**5.**
**Dividend** _____

**Divisor** _____

**Quotient** _____

## Objective 3    Divide 0 by a number
## Objective 4  Recognize that a number cannot be divided by 0.

*Divide. If the division is not possible, write* **"undefined."**

6.    $12\overline{)0}$

**6.** _____

7.    $0\overline{)72}$

**7.** _____

8.    $\dfrac{0}{6}$

**8.** _____

9.    $\dfrac{7}{0}$

**9.** _____

10.    $0 \div 15$

**10.** _____

11.    $9 \div 0$

**11.** _____

**Objective 5   Divide a number by itself**
**Objective 6  Divide a number by 1.**

*Divide.*

12.   $18 \div 18$                                        12. _____

13.   $1\overline{)38}$                                      13. _____

**Objective 7   Use short division.**

*Divide by using short division.*

14.   $2\overline{)84}$                                      14. _____

15.   $724 \div 5$                                       15. _____

16.   $\dfrac{651}{9}$                                       16. _____

17.   $8\overline{)1135}$                                    17. _____

18.   $984 \div 6$                                       18. _____

19.   $\dfrac{512}{3}$                                       19. _____

**Objective 8   Use multiplication to check the answer to a division problem.**

*Use multiplication to check each answer. If an answer is incorrect, find the correct answer.*

20.   $6\overline{)9137}$ with quotient $1522$ R4          20. _____

21.   $3852 \div 4 = 963$                                21. _____

22.   $\dfrac{8621}{3} = 2873 \ \text{R} \ 2$                            22. _____

**23.**   $7\overline{)40,698}$ ^(4814)

**23.** _____

**24.**   $\dfrac{18,150}{3} = 650$

**24.** _____

**25.**   $20,351 \div 6 = 3391 \text{ R } 5$

**25.** _____

## Objective 9   Use tests for divisibility.

*Determine if the following numbers are divisible by 2, 3, 5, or 10 Write* **yes** *or* **no**.

**26.**   50

**26. 2:** _____

      **3:** _____

      **5:** _____

     **10:** _____

**27.**   897

**27. 2:** _____

      **3:** _____

      **5:** _____

     **10:** _____

**28.**   908

**28. 2:** _____

      **3:** _____

      **5:** _____

     **10:** _____

**29.**   6205

**29. 2:** _____

      **3:** _____

      **5:** _____

     **10:** _____

**30.**   32,175

**30. 2:** _____

      **3:** _____

      **5:** _____

     **10:** _____

# Chapter 1 WHOLE NUMBERS

### 1.6    Long Division

| Learning Objectives |
| --- |
| 1    Do long division. |
| 2    Divide numbers ending in 0 by numbers ending in 0. |
| 3    Use multiplication to check division answers. |

### Key Terms

Use the vocabulary terms listed below to complete each statement in exercises 1–5.

**long division        dividend        divisor        quotient        remainder**

1.  In the problem $751 \div 23 = 32 \text{ R } 15$, 751 is called the _____.

2.  In the problem $751 \div 23 = 32 \text{ R } 15$, 15 is called the _____.

3.  In the problem $751 \div 23 = 32 \text{ R } 15$, 23 is called the _____.

4.  In the problem $751 \div 23 = 32 \text{ R } 15$, 32 is called the _____.

5.  _____ is a method of dividing a number by a divisor with more than one digit.

### Objective 1    Do long division.

*Divide using long division. Check each answer.*

1.  $32\overline{)2624}$                         1. _____

2.  $29\overline{)9396}$                         2. _____

3.  $42\overline{)3234}$                         3. _____

**4.** $23\overline{)1587}$

**4.** _____

**5.** $53\overline{)5406}$

**5.** _____

**6.** $37\overline{)4215}$

**6.** _____

**7.** $89\overline{)7649}$

**7.** _____

**8.** $56\overline{)9314}$

**8.** _____

**9.** $94\overline{)29,047}$

**9.** _____

**10.** $71\overline{)412,794}$

**10.** _____

**11.** $28\overline{)177,919}$

**11.** _____

**12.** $86\overline{)8,473,758}$

**12.** _____

**13.** $205\overline{)6,680,335}$

**13.** _____

**14.** $327\overline{)98,413,712}$

**14.** _____

**15.** $657\overline{)429,700}$

**15.** _____

**16.** $732\overline{)4,268,292}$

**16.** _____

**Objective 2     Divide numbers ending in 0 by numbers ending in 0.**

*Divide.*

17.     $80\overline{)560}$                                    17. _____

18.     $400\overline{)6000}$                               18. _____

19.     $2000\overline{)12,000}$                          19. _____

20.     $800\overline{)10,400}$                          20. _____

21.     $910\overline{)38,220}$                          21. _____

22.     $750\overline{)25,500}$                          22. _____

23.     $1200\overline{)960,000}$                     23. _____

**Objective 3     Use multiplication to check division answers.**

*Check each answer. If an answer is incorrect, give the correct answer.*

24.     $37\overline{)3235}$   87 R16                            24. _____

25.     $89\overline{)5790}$   65 R5                             25. _____

**26.**
$$74 \overline{)25{,}621} \quad \begin{array}{c} 346 \ \text{R}18 \end{array}$$

**26.** _____

**27.**
$$103 \overline{)4658} \quad \begin{array}{c} 44 \ \text{R}22 \end{array}$$

**27.** _____

**28.**
$$205 \overline{)47{,}538} \quad \begin{array}{c} 231 \ \text{R}183 \end{array}$$

**28.** _____

**29.**
$$318 \overline{)94{,}207} \quad \begin{array}{c} 297 \ \text{R}79 \end{array}$$

**29.** _____

**30.**
$$428 \overline{)196{,}883} \quad \begin{array}{c} 400 \ \text{R}30 \end{array}$$

**30.** _____

## Chapter 1 WHOLE NUMBERS

### 1.7    Rounding Whole Numbers

| Learning Objectives |
| --- |
| 1    Locate the place to which a number is to be rounded. |
| 2    Round numbers. |
| 3    Round numbers to estimate an answer. |
| 4    Use front end rounding to estimate an answer. |

### Key Terms

Use the vocabulary terms listed below to complete each statement in exercises 1–3.

**rounding      estimate      front end rounding**

1.    _____ is rounding to the highest possible place so that all the digits become zeros except the first one.

2.    In order to find a number that is close to the original number, but easier to work with, use a process called _____.

3.    _____ to find an answer close to the exact answer.

### Objective 1    Locate the place to which a number is to be rounded.

*Locate the place to which the number is rounded by underlining the appropriate digit.*

| 1. | 257,301 | Nearest ten | 1. _____ |
| 2. | 1037 | Nearest hundred | 2. _____ |
| 3. | 645,371 | Nearest ten-thousand | 3. _____ |
| 4. | 39,943,712 | Nearest million | 4. _____ |

### Objective 2    Round numbers.

*Round each number as indicated.*

| 5. | 7863 to the nearest hundred | 5. _____ |
| 6. | 1382 to the nearest ten | 6. _____ |
| 7. | 18,211 to the nearest hundred | 7. _____ |
| 8. | 9348 to the nearest hundred | 8. _____ |

**9.**    8398 to the nearest hundred            **9.** _____

**10.**    41,099 to the nearest hundred        **10.** _____

**11.**    51,803 to the nearest thousand       **11.** _____

**12.**    16,968 to the nearest hundred        **12.** _____

**13.**    53,595 to the nearest hundred        **13.** _____

**14.**    476,943 to the nearest ten-thousand   **14.** _____

**15.**    576,295 to the nearest hundred-thousand   **15.** _____

**16.**    14,823,307 to the nearest million     **16.** _____

**Objective 3    Round numbers to estimate an answer.**

*Estimate each answer by rounding to the nearest ten. Then find the exact answer.*

**17.**    37
          24
          58
      + 91

**17.**
**Estimate** _____

**Exact** _____

**18.**    19
          87
          35
      + 20

**18.**
**Estimate** _____

**Exact** _____

**19.**    69
      − 42

**19.**
**Estimate** _____

**Exact** _____

**20.**      88
          – 52

20.

**Estimate** _____

**Exact** _____

*Estimate each answer by rounding to the nearest hundred. Then find the exact answer.*

**21.**      276
          312
          174
        + 936

21.

**Estimate** _____

**Exact** _____

**22.**      419
          188
          324
        + 194

22.

**Estimate** _____

**Exact** _____

**23.**      971
          – 382

23.

**Estimate** _____

**Exact** _____

**24.**      815
          – 678

24.

**Estimate** _____

**Exact** _____

**25.**      912
          × 784

25.

**Estimate** _____

**Exact** _____

**26.**      876
          × 141

26.

**Estimate** _____

**Exact** _____

## Objective 4    Use front end rounding to estimate an answer.

*Estimate each answer using front end rounding. Then find the exact answer.*

**27.**        571
                42
               215
           + 2452

**27.**

**Estimate** _____

**Exact** _____

**28.**        313
            −   49

**28.**

**Estimate** _____

**Exact** _____

**29.**        980
            ×   37

**29.**

**Estimate** _____

**Exact** _____

**30.**        437
            ×   29

**30.**

**Estimate** _____

**Exact** _____

# Chapter 1 WHOLE NUMBERS

### 1.8     Exponents, Roots, and Order of Operations

| **Learning Objectives** |
| --- |
| 1     Identify an exponent and a base. |
| 2     Find the square root of a number. |
| 3     Use the order of operations. |

### Key Terms

Use the vocabulary terms listed below to complete each statement in exercises 1–3.

**square root          perfect square          order of operations**

1.  For problems or expressions with more than one operation, the _____ tells what to do first, second, and so on, to obtain the correct answer.

2.  The _____ of a whole number is the number that can be multiplied by itself to produce the given number.

3.  A _____ is a number that is the square of a whole number.

### Objective 1     Identify an exponent and a base.

*Identify the exponent and the base, then simplify each expression.*

1.  $7^2$

    **1. Exponent _____**

    **Base _____**

    **Expression _____**

2.  $2^7$

    **2. Exponent _____**

    **Base _____**

    **Expression _____**

3.  $8^3$

    **3. Exponent _____**

    **Base _____**

    **Expression _____**

**4.**    $10^4$ 

                                       **4. Exponent** _____

                                       **Base** _____

                                       **Expression** _____

## Objective 2    Find the square root of a number.

*Find each square root.*

**5.**    $\sqrt{16}$                                       **5.** _____

**6.**    $\sqrt{64}$                                       **6.** _____

**7.**    $\sqrt{121}$                                    **7.** _____

**8.**    $\sqrt{169}$                                    **8.** _____

**9.**    $\sqrt{225}$                                    **9.** _____

*Fill in each blank.*

**10.**    $18^2 =$ \_\_\_\_\_ so $\sqrt{\rule{1.5em}{0pt}} = 18$          **10.** _____

**11.**    $50^2 =$ \_\_\_\_\_ so $\sqrt{\rule{1.5em}{0pt}} = 50$          **11.** _____

**12.**    $25^2 =$ \_\_\_\_\_ so $\sqrt{\rule{1.5em}{0pt}} = 25$          **12.** _____

**13.**    $20^2 =$ \_\_\_\_\_ so $\sqrt{400} =$ \_\_\_\_\_          **13.** _____

**14.**    $36^2 =$ \_\_\_\_\_ so $\sqrt{1296} =$ \_\_\_\_\_        **14.** _____

## Objective 3    Use the order of operations.

*Simplify each expression using the order of operations.*

**15.**    $6^2 + 5 - 2$                                **15.** _____

**16.** $2^4 + 3 \cdot 4 - 5$

**16.** _____

**17.** $6 \cdot 5 - 5 \div 0$

**17.** _____

**18.** $8 \cdot 9 \div 12$

**18.** _____

**19.** $9 \cdot 7 - 3 \cdot 12$

**19.** _____

**20.** $7 - 25 \div 5$

**20.** _____

**21.** $6 \cdot 3^2 + 0 \div 6$

**21.** _____

**22.** $8 \cdot 5 - 12 \div (2 \cdot 3 - 6)$

**22.** _____

**23.** $4 \cdot 3 + 8 \cdot 5 - 7$

**23.** _____

**24.** $4 \cdot (9 - 7) + 3 \cdot 8$

**24.** _____

**25.** $2^3 \cdot 3^2 + 5(3) \div 5$

**25.** _____

**26.**  $6 \cdot \sqrt{144} - 6 \cdot 8$

**26.** _____

**27.**  $42 \div 6 + 3 \cdot \sqrt{49}$

**27.** _____

**28.**  $2 \cdot \sqrt{121} - 2 \div \sqrt{4} + \left(14 - 2 \cdot 7\right) \div 4$

**28.** _____

**29.**  $25 \div 5 \cdot 3 \cdot 9 \div \left(14 - 11\right)$

**29.** _____

**30.**  $3^2 \cdot \sqrt{36} \div \sqrt{81} \div 3 + 2 \cdot 3 - 2$

**30.** _____

# Chapter 1 WHOLE NUMBERS

## 1.9    Reading Pictographs, Bar Graphs, and Line Graphs

| **Learning Objectives** |
| --- |
| 1      Read and understand a pictograph. |
| 2      Read and understand a bar graph. |
| 3      Read and understand a line graph. |

## Key Terms

Use the vocabulary terms listed below to complete each statement in exercises 1–3.

       **pictograph**         **bar graph**          **line graph**

**1.**   A _____ is used to display a trend.

**2.**   A graph that uses pictures or symbols to display information is called a
     _____.

**3.**   A _____ uses bars of various heights to show quantity.

## Objective 1    Read and understand a pictograph.

*The pictograph shows the amount of sales tax in various states. Use the pictograph to answer exercises 1–5.*

State Sales Tax

Source: Federation of Tax Administrators

**1.**   Which state shown in the pictograph charges the least sales tax?

1. _____

**2.**   Which state has a sales tax of 5%?

2. _____

**3.**   According to the pictograph, which state has the greatest sales tax?

3. _____

**4.** Which state has a sales tax of 4%?                4. _____

**5.** By about how much does Minnesota's sales tax
exceed Utah's sales tax?                                5. _____

*The pictograph shows the number of male and female students in the school's language
clubs. Use the pictograph to answer exercises 6–10.*

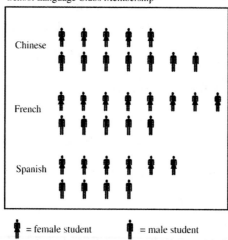

School Language Clubs Membership

= female student          = male student

**6.** Which language club has the largest number of
members?                                               6. _____

**7.** How many more female students are there than male
students in the French club?                           7. _____

**8.** What is the total number of members in the Chinese
club?                                                  8. _____

**9.** Which language club has the least number of
members?                                               9. _____

**10.** How many more female students are in the French
and Spanish clubs combined than male students?         10. _____

## Objective 2    Read and understand a bar graph.

*The bar graph shows the number of ice cream cones of different flavors that were sold at a barbecue. Use the bar graph to answer exercises 11–15.*

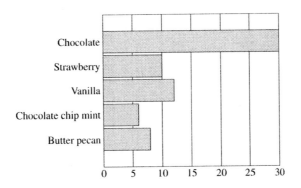

11.    How many strawberry cones were sold?                    11. _____

12.    Which flavor had the fewest sales?                       12. _____

13.    How many more vanilla cones were sold than butter        13. _____
       pecan cones?

14.    How many more chocolate cones were sold than             14. _____
       vanilla and strawberry combined?

15.    How many strawberry and butter pecan cones were          15. _____
       sold in total?

*The bar graph shows the enrollment by gender in each class at a small college. Use the bar graph to answer questions 16–20.*

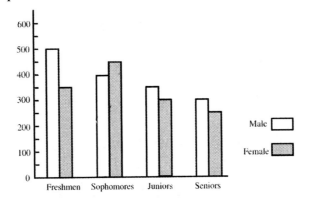

16.    How many more male freshmen are there than               16. _____
       female seniors?

**17.** Find the total number of students enrolled.        **17.** _____

**18.** How many more sophomores are there than juniors?        **18.** _____

**19.** Which class has the greatest difference between male students and female students?        **19.** _____

**20.** Which class has more female students than male students?        **20.** _____

**Objective 3    Read and understand a line graph.**

*The line graph shows the net sales for Ajax Systems from 2003 to 2007. Use the line graph to answer exercises 21–25.*

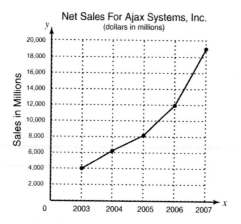

**21.** What trend or pattern is shown in the graph?        **21.** _____

**22.** Approximately what were the net sales in 2003?        **22.** _____

**23.** Which year had the largest increase over the previous year?        **23.** _____

**24.** Which year had the highest net sales?        **24.** _____

**25.** Between which two years was the increase smallest?        **25.** _____

*The line graph shows the annual sales for two different stores from 1996 to 2000. Use the line graph to answer exercises 26–30.*

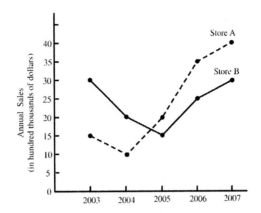

26.   In which years did the sales of store A exceed the          26. _____
      sales of store B?


27.   Which year showed the least difference between the          27. _____
      sales of store A and the sales of store B?


28.   Which year showed the greatest difference between           28. _____
      the sales of store A and the sales of store B?


29.   What was the difference in annual sales between             29. _____
      store A and store B in 1996?


30.   Between which two years was the increase for both           30. _____
      stores greatest?

## Chapter 1 WHOLE NUMBERS

### 1.10    Solving Application Problems

| **Learning Objectives** |
| --- |
| 1     Find indicator words in application problems. |
| 2     Solve application problems. |
| 3     Estimate an answer. |

### Key Terms

Use the vocabulary terms listed below to complete each statement in exercises 1–5.

| indicator words | sum | difference | product | quotient |
| --- | --- | --- | --- | --- |
| increased by | fewer | times | per | |

1. Words in a problem that indicate the necessary operations are

    _____.

2. _____ and _____ are indicator words for addition.

3. _____ and _____ are indicator words for multiplication.

4. _____ and _____ are indicator words for division.

5. _____ and _____ are indicator words for subtraction.

### Objective 1    Find indicator words in application problems.

*Write the operation determined by each of the following.*

1. decreased by                         1. _____

2. more than                             2. _____

3. twice                                    3. _____

4. goes into                             4. _____

5. loss of                                 5. _____

## Objective 2   Solve application problems.

*Solve each application problem.*

6.  A truck weights 8950 pounds when empty. After being loaded with firewood, it weighs 17,180 pounds. What is the weight of the firewood?

6. _____

7.  How many 3-inch strips of leather can be cut from a piece of leather 1 foot wide? (Hint: 1 foot = 12 inches.)

7. _____

8.  If there are 43,560 square feet in an acre, how many square feet are there in 5 acres?

8. _____

9.  Travel Rent-A-Car owns 365 compact cars, 438 full-sized cars, 125 luxury cars, and 83 vans and trucks. How many vehicles does it have in all?

9. _____

10. Amanda Raymond owes $5520 on a loan. Find her monthly payment if the loan is paid off in 48 months.

10. _____

11. Total receipts at a concert were $191,800. Each ticket cost $28. How many people attended the concert?

11. _____

12. Two sisters share a legal bill of $1903. One sister pays $954 toward the bill. How much must the other sister pay?

12. _____

**13.** A biology class found 14 deer in one area, 158 in another, and 417 in a third. How many deer did the class find?

**13.** _____

**14.** Baseball uniforms cost $79 each. Find the cost of 23 uniforms.

**14.** _____

**15.** The number of gallons of water polluted each day in an industrial area is 219,530. How many gallons are polluted each year? (Use a 365-day year.)

**15.** _____

**16.** Shari bought 3 books costing $12, $17, and $18 each. She paid with a $50 bill. How much change will she receive?

**16.** _____

**17.** A new car costs $11,350 before a trade-in. The car can be paid off in 36 monthly payments of $209 each after the trade-in. Find the amount of the trade-in.

**17.** _____

**18.** Blue Bird leader, Barbara Walton, estimates that each of her Blue Birds will eat 2 cookies while she and her assistant, Lana Meehan, will eat 3 cookies each. If she expects 15 Blue Birds and her assistant at the meeting, how many cookies will she need?

**18.** _____

**19.** Edward Biondi has $3117 in his checking account. If
he pays $340 for tires, $725 for equipment repairs,
and $198 for fuel and oil, find the balance remaining
in his account.

**19.** _____

**20.** Rodney Guess owns 55 acres of land which he
leases to an alfalfa farmer for $150 per acre per year.
If property taxes are $28 per acre per year, find the
total amount he has left after taxes are paid.

**20.** _____

**Objective 3    Estimate an answer.**

*First use front end rounding to estimate the answer. Then find the exact answer.*

**21.** A bus traveled 605 miles at 55 miles per hour. How
long did the trip take?

**21.**
**Estimate** _____

**Exact** _____

**22.** Liz Skinner has $3712 in her checking account.
After writing a check of $887 for tuition and parking
fees, how much remains in her account?

**22.**
**Estimate** _____

**Exact** _____

**23.** Lori Knight knows that her car gets 36 miles per
gallon in town. How many miles can she travel on
26 gallons?

**23.**
**Estimate** _____

**Exact** _____

**24.** Ski Mart offers a set of skis at a sale price of $219.
If the sale price gives a savings of $56 off the
original price, what is the original price of the skis?

**24.**
**Estimate** _____

**Exact** _____

**25.** If there are 43,560 square feet in an acre, how many square feet are there in 5 acres?

**25.**
**Estimate**_____

**Exact** _____

**26.** If 560 stamps are divided evenly among 16 collectors, how many stamps will each receive?

**26.**
**Estimate**_____

**Exact** _____

**27.** A room measures 18 feet by 12 feet. If carpeting costs $23 per square yard, find the total cost for carpeting the room.
(Hint: one square yard = 3 feet $\times$ 3 feet.)

**27.**
**Estimate**_____

**Exact** _____

**28.** The Top Hat Grille finds that it needs five pounds of hamburger to make 35 servings of chili. How many pounds of hamburger are needed to make 182 servings of chili?

**28.**
**Estimate**_____

**Exact** _____

**29.** Jerri Taft's vending machine company had 325 machines on hand at the beginning of the month. At different times during the month, machines were distributed to new locations: 37 machines were taken at one time, then 24 machines, and then 81 machines. During the same month additional machines were returned: 16 machines were returned at one time, then 39 machines, and then 110 machines. How many machines were on hand at the end of the month?

**29.**
**Estimate**_____

**Exact** _____

**30.** Diana Ditka spent $286 on tuition, $137 on books, and $32 on supplies. If this money is withdrawn from her checking account, which had a balance of $723, what is her new balance.

**30.**
**Estimate**_____

**Exact** _____

Name:                 Date:

Instructor:          Section:

# Chapter 2 MULTIPLYING AND DIVIDING FRACTIONS

### 2.1    Basics of Fractions

| Learning Objectives |
| --- |
| 1     Use a fraction to show which part of a whole is shaded. |
| 2     Identify the numerator and denominator. |
| 3     Identify proper and improper fractions. |

### Key Terms

Use the vocabulary terms listed below to complete each statement in exercises 1–4.

**numerator**        **denominator**        **proper fraction**

**improper fraction**

1. A fraction whose numerator is larger than its denominator is called an
   _____.

2. In the fraction $\frac{2}{9}$, the 2 is the _____.

3. A fraction whose denominator is larger than its numerator is called a
   _____.

4. The _____ of a fraction shows the number of equal
   parts in a whole.

### Objective 1    Use a fraction to show which part of a whole is shaded.

*Write the fractions that represent the shaded and unshaded portions of each figure.*

1.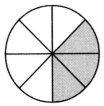

   **1. Shaded** _____

        **Unshaded** _____

2.

   **2. Shaded** _____

        **Unshaded** _____

3.

   **3. Shaded** _____

        **Unshaded** _____

**4.**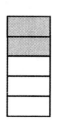

**4. Shaded** _____

   **Unshaded** _____

**5.**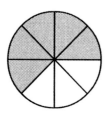

**5. Shaded** _____

   **Unshaded** _____

**6.**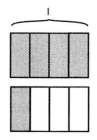

**6. Shaded** _____

   **Unshaded** _____

**7.**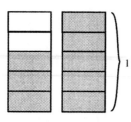

**7. Shaded** _____

   **Unshaded** _____

**8.**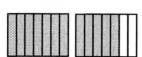

**8. Shaded** _____

   **Unshaded** _____

**9.**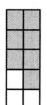

**9. Shaded** _____

   **Unshaded** _____

**10.**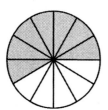

**10. Shaded** _____

   **Unshaded** _____

## Objective 2    Identify the numerator and denominator.

*Identify the numerator and denominator.*

11.   $\dfrac{4}{3}$

12.   $\dfrac{2}{5}$

13.   $\dfrac{8}{11}$

14.   $\dfrac{11}{8}$

15.   $\dfrac{19}{50}$

16.   $\dfrac{112}{5}$

17.   $\dfrac{19}{8}$

18.   $\dfrac{98}{13}$

19.   $\dfrac{157}{12}$

**11.**
**Numerator**_____

**Denominator** _____

**12.**
**Numerator**_____

**Denominator** _____

**13.**
**Numerator**_____

**Denominator** _____

**14.**
**Numerator**_____

**Denominator** _____

**15.**
**Numerator**_____

**Denominator** _____

**16.**
**Numerator**_____

**Denominator** _____

**17.**
**Numerator**_____

**Denominator** _____

**18.**
**Numerator**_____

**Denominator** _____

**19.**
**Numerator**_____

**Denominator** _____

**20.**  $\dfrac{14}{195}$

**20.**

**Numerator** _____

**Denominator** _____

### Objective 3    Identify proper and improper fractions.

*Write whether each fraction is **proper** or **improper**.*

**21.**  $\dfrac{9}{7}$

**21.** _____

**22.**  $\dfrac{5}{12}$

**22.** _____

**23.**  $\dfrac{7}{15}$

**23.** _____

**24.**  $\dfrac{17}{11}$

**24.** _____

**25.**  $\dfrac{4}{19}$

**25.** _____

**26.**  $\dfrac{11}{7}$

**26.** _____

**27.**  $\dfrac{18}{18}$

**27.** _____

**28.**  $\dfrac{7}{12}$

**28.** _____

**29.**  $\dfrac{3}{4}$

**29.** _____

**30.**  $\dfrac{10}{10}$

**30.** _____

# Chapter 2 MULTIPLYING AND DIVIDING FRACTIONS

## 2.2 Mixed Numbers

| Learning Objectives |
| --- |
| 1      Identify mixed numbers. |
| 2      Write mixed numbers as improper fractions. |
| 3      Write improper fractions as mixed numbers. |

### Key Terms

Use the vocabulary terms listed below to complete each statement in exercises 1–4.

**mixed number      improper fraction      proper fraction**

**whole numbers**

1. The fraction $\dfrac{5}{8}$ is an example of a _____.

2. A _____ includes a fraction and a whole number written together.

3. A mixed number can be rewritten as an _____.

4. 0, 1, 2, 3, … are _____.

### Objective 1      Identify mixed numbers.

*List the mixed numbers in each group.*

1. $2\dfrac{1}{2}, \dfrac{3}{5}, 1\dfrac{1}{6}, \dfrac{3}{4}$          1. _____

2. $\dfrac{3}{8}, 5\dfrac{2}{3}, \dfrac{7}{4}, 3\dfrac{1}{2}$          2. _____

3. $\dfrac{8}{7}, \dfrac{10}{10}, \dfrac{2}{3}, \dfrac{0}{5}$          3. _____

4. $\dfrac{9}{9}, 3\dfrac{1}{2}, 10\dfrac{1}{3}, \dfrac{8}{2}, \dfrac{7}{9}$          4. _____

5. $\dfrac{6}{3}, 4\dfrac{3}{4}, \dfrac{10}{10}, \dfrac{1}{3}, \dfrac{0}{8}$          5. _____

## Objective 2    Write mixed numbers as improper fractions.

*Write each mixed number as an improper fraction.*

**6.** $2\frac{7}{8}$

**6.** _____

**7.** $1\frac{5}{6}$

**7.** _____

**8.** $2\frac{4}{5}$

**8.** _____

**9.** $5\frac{4}{7}$

**9.** _____

**10.** $1\frac{3}{4}$

**10.** _____

**11.** $6\frac{1}{4}$

**11.** _____

**12.** $4\frac{2}{3}$

**12.** _____

**13.** $7\frac{1}{2}$

**13.** _____

**14.** $2\frac{7}{11}$

**14.** _____

**15.** $5\frac{3}{7}$

**15.** _____

**16.** $6\frac{2}{3}$

**16.** _____

**17.** $8\frac{7}{9}$

**17.** _____

**18.** $13\frac{3}{9}$

**18.** _____

**Objective 3    Write improper fractions as mixed numbers.**

*Write each improper fraction as a mixed number.*

19.  $\dfrac{14}{9}$

19. _____

20.  $\dfrac{20}{7}$

20. _____

21.  $\dfrac{29}{9}$

21. _____

22.  $\dfrac{26}{7}$

22. _____

23.  $\dfrac{21}{5}$

23. _____

24.  $\dfrac{41}{9}$

24. _____

25.  $\dfrac{25}{9}$

25. _____

26.  $\dfrac{29}{4}$

26. _____

27.  $\dfrac{92}{3}$

27. _____

28.  $\dfrac{211}{11}$

28. _____

29.  $\dfrac{749}{17}$

29. _____

30.  $\dfrac{2573}{11}$

30. _____

# Chapter 2 MULTIPLYING AND DIVIDING FRACTIONS

## 2.3     Factors

| Learning Objectives |
|---|
| 1     Find factors of a number. |
| 2     Identify prime numbers |
| 3     Find prime factorizations. |

## Key Terms

Use the vocabulary terms listed below to complete each statement in exercises 1–5.

**factors**          **composite number**        **prime number**

**factorizations**                          **prime factorization**

1.    The numbers that can be multiplied to give a specific number (product) are _____ of that number.

2.    A _____ has at least one factor other than itself and 1.

3.    In a _____ every factor is a prime number.

4.    The factors of a _____ are itself and 1.

5.    Numbers that are multiplied to give a product are _____.

## Objective 1     Find factors of a number.

*Find all the factors of each number.*

1.    7                                        1.  _____

2.    12                                       2.  _____

3.    49                                       3.  _____

4.    15                                       4.  _____

5.    10                                       5.  _____

6.    36                                       6.  _____

**7.** 25                                **7.** _____

**8.** 24                                **8.** _____

**9.** 30                                **9.** _____

**10.** 72                               **10.** _____

## Objective 2    Identify prime numbers.

*Write whether each number is* **prime**, **composite**, *or* **neither**.

**11.** 1                                **11.** _____

**12.** 5                                **12.** _____

**13.** 15                               **13.** _____

**14.** 11                               **14.** _____

**15.** 24                               **15.** _____

**16.** 45                               **16.** _____

**17.** 2                                **17.** _____

**18.** 31                               **18.** _____

**19.** 29                               **19.** _____

**20.** 38                               **20.** _____

## Objective 3    Find prime factorizations.

*Find the prime factorization of each number. Write the answer with exponents when repeated factors appear.*

**21.** 12                               **21.** _____

**22.** 27                              **22.** _____

**23.** 28                              **23.** _____

**24.** 42                              **24.** _____

**25.** 24                              **25.** _____

**26.** 72                              **26.** _____

**27.** 108                             **27.** _____

**28.** 160                             **28.** _____

**29.** 450                             **29.** _____

**30.** 171                             **30.** _____

## Chapter 2 MULTIPLYING AND DIVIDING FRACTIONS

### 2.4     Writing a Fraction in Lowest Terms

**Learning Objectives**
| | |
|---|---|
| 1 | Tell whether a fraction is written in lowest terms. |
| 2 | Write a fraction in lowest terms using common factors. |
| 3 | Write a fraction in lowest terms using prime factors. |
| 4 | Determine whether two fractions are equivalent. |

**Key Terms**

Use the vocabulary terms listed below to complete each statement in exercises 1–3.

> **equivalent fractions**      **common factor**      **lowest terms**

1. A fraction is written in _____ when its numerator and denominator have no common factor other than 1.

2. A _____ is a number that can be divided into two or more whole numbers.

3. Two fractions are _____ when they represent the same portion of a whole.

### Objective 1     Tell whether a fraction is written in lowest terms.

*Write whether or not each fraction is in lowest terms. Write* **yes** *or* **no**.

1. $\dfrac{4}{12}$                    1. _____

2. $\dfrac{3}{7}$                     2. _____

3. $\dfrac{12}{18}$                    3. _____

4. $\dfrac{13}{17}$                    4. _____

5. $\dfrac{7}{19}$                     5. _____

6. $\dfrac{3}{39}$                     6. _____

## Objective 2    Write a fraction in lowest terms using common factors.

*Write each fraction in lowest terms.*

7.    $\dfrac{4}{12}$                                     7. _____

8.    $\dfrac{14}{49}$                                   8. _____

9.    $\dfrac{8}{36}$                                     9. _____

10.    $\dfrac{26}{39}$                                  10. _____

11.    $\dfrac{28}{98}$                                  11. _____

12.    $\dfrac{30}{42}$                                  12. _____

13.    $\dfrac{12}{88}$                                  13. _____

14.    $\dfrac{16}{56}$                                  14. _____

## Objective 3    Write a fraction in lowest terms using prime factors.

*Write the numerator and denominator of each fraction as a product of prime factors and divide by the common factors. Then write the fraction in lowest terms.*

**15.**    $\dfrac{63}{84}$

**15.** _____

**16.**    $\dfrac{28}{56}$

**16.** _____

**17.**    $\dfrac{180}{210}$

**17.** _____

**18.**    $\dfrac{72}{90}$

**18.** _____

**19.**    $\dfrac{36}{54}$

**19.** _____

**20.**    $\dfrac{71}{142}$

**20.** _____

**21.**    $\dfrac{75}{500}$

**21.** _____

**22.**    $\dfrac{96}{132}$

**22.** _____

## Objective 4    Determine whether two fractions are equivalent.

*Determine whether each pair of fractions is* **equivalent** *or* **not equivalent**.

**23.**    $\dfrac{3}{7}$ and $\dfrac{6}{14}$

23. _____

**24.**    $\dfrac{2}{3}$ and $\dfrac{10}{15}$

24. _____

**25.**    $\dfrac{6}{21}$ and $\dfrac{3}{7}$

25. _____

**26.**    $\dfrac{8}{16}$ and $\dfrac{15}{20}$

26. _____

**27.**    $\dfrac{9}{12}$ and $\dfrac{6}{8}$

27. _____

**28.**    $\dfrac{12}{28}$ and $\dfrac{18}{42}$

28. _____

**29.**    $\dfrac{20}{24}$ and $\dfrac{15}{31}$

29. _____

**30.**    $\dfrac{6}{12}$ and $\dfrac{8}{16}$

30. _____

# Chapter 2 MULTIPLYING AND DIVIDING FRACTIONS

### 2.5    Multiplying Fractions

| **Learning Objectives** |
| :--- |
| 1    Multiply fractions. |
| 2    Use a multiplication shortcut. |
| 3    Multiply a fraction and a whole number. |
| 4    Find the area of a rectangle. |

## Key Terms

Use the vocabulary terms listed below to complete each statement in exercises 1–4.

**multiplication shortcut**        **numerator**            **denominator**

**common factor**

1.    A _____ can be divided into two or more whole numbers.

2.    The number below the fraction bar in a fraction is called the _____.

3.    When multiplying fractions, the process of dividing a numerator and denominator by a common factor can be used as a _____.

4.    The number above the fraction bar in a fraction is called the _____.

### Objective 1    Multiply factions.

*Multiply. Write answers in lowest terms.*

1.    $\dfrac{5}{9} \cdot \dfrac{7}{6}$

1.    _____

2.    $\dfrac{4}{7} \cdot \dfrac{3}{5}$

2.    _____

3.    $\dfrac{5}{6} \cdot \dfrac{11}{4}$

3.    _____

4.    $\dfrac{9}{10} \cdot \dfrac{3}{2}$

4.    _____

5.    $\dfrac{3}{4} \cdot \dfrac{5}{6} \cdot \dfrac{2}{3}$

5.    _____

Name:                                    Date:
Instructor:                              Section:

6.  $\dfrac{1}{9} \cdot \dfrac{2}{3} \cdot \dfrac{5}{6}$                      6. _____

7.  $\dfrac{3}{8} \cdot \dfrac{1}{4} \cdot \dfrac{1}{9}$                      7. _____

## Objective 2   Use a multiplication shortcut.

*Use the multiplication shortcut to find each product. Write the answer in lowest terms.*

8.  $\dfrac{7}{6} \cdot \dfrac{3}{14}$                         8. _____

9.  $\dfrac{4}{9} \cdot \dfrac{15}{16}$                        9. _____

10. $\dfrac{3}{5} \cdot \dfrac{25}{27}$                        10. _____

11. $\dfrac{11}{4} \cdot \dfrac{8}{33}$                        11. _____

12. $\dfrac{5}{6} \cdot \dfrac{4}{35}$                         12. _____

13. $\dfrac{3}{4} \cdot \dfrac{5}{9} \cdot \dfrac{2}{5}$                13. _____

14. $\dfrac{3}{8} \cdot \dfrac{4}{9} \cdot \dfrac{15}{6}$               14. _____

15. $\dfrac{25}{35} \cdot \dfrac{14}{30} \cdot \dfrac{3}{7}$            15. _____

## Objective 3     Multiply a fraction and a whole number.

*Multiply. Write the answer in lowest terms. Change the answer to a whole or mixed number where possible.*

**16.**    $6 \cdot \dfrac{7}{300}$                                       **16.** _____

**17.**    $\dfrac{4}{250} \cdot 50$                                       **17.** _____

**18.**    $27 \cdot \dfrac{7}{54}$                                       **18.** _____

**19.**    $49 \cdot \dfrac{6}{7}$                                       **19.** _____

**20.**    $21 \cdot \dfrac{3}{7} \cdot \dfrac{7}{9}$                                  **20.** _____

**21.**    $\dfrac{9}{26} \cdot \dfrac{39}{18} \cdot 12$                               **21.** _____

**22.**    $200 \cdot \dfrac{7}{50} \cdot \dfrac{5}{28}$                             **22.** _____

**23.**    $\dfrac{21}{520} \cdot 13 \cdot \dfrac{20}{7}$                            **23.** _____

## Objective 4     Find the area of a rectangle.

*Find the area of each rectangle.*

**24.**    Length: $\frac{2}{3}$ yard, width: $\frac{1}{2}$ yard            **24.** _____

**25.**    $\frac{3}{4}$ meter, width: $\frac{1}{2}$ meter                                  **25.** _____

**26.**    Length: $\frac{5}{3}$ yards, width: $\frac{3}{2}$ yards                   **26.** _____

**27.**    Length: $\frac{9}{16}$ meter, width: $\frac{5}{6}$ meter                 **27.** _____

**28.**    Length: $\frac{7}{16}$ inch, width: $\frac{3}{16}$ inch                   **28.** _____

*Solve each application problem.*

**29.**    A desk is $\frac{2}{3}$ yard by $\frac{5}{6}$ yard. Find its area.            **29.** _____

**30.**    A wading pool is $\frac{5}{4}$ yards by $\frac{5}{9}$ yard. Find its area      **30.** _____

Name:                              Date:

Instructor:                  Section:

# Chapter 2 MULTIPLYING AND DIVIDING FRACTIONS

## 2.6      Applications of Multiplication

| **Learning Objectives** |
| --- |
| 1      Solve fraction application problems using multiplication. |

## Key Terms

Use the vocabulary terms listed below to complete each statement in exercises 1–3.

**reciprocal**            **product**       **indicator words**

1. The words "times" and "double" are _____ for multiplication.

2. In the problem $51 \times 3 = 153$, 153 is called the _____.

3. Two numbers are _____ of each other if their product is 1.

## Objective 1     Solve fraction application problems using multiplication.

*Solve each application problem.*

1. A bookstore sold 2800 books, $\frac{3}{5}$ of which were paperbacks. How many paperbacks were sold?

     1. _____

2. A store sells 3750 items, of which $\frac{2}{15}$ are classified as junk food. How many of the items are junk food?

     2. _____

3. Sara needs $2500 to go to school for one year. She earns $\frac{3}{5}$ of this amount in the summer. How much does she earn in the summer?

     3. _____

4. Lani paid $120 for textbooks this term. Of this amount, the bookstore kept $\frac{1}{4}$. How much did the bookstore keep?

     4. _____

**5.**  Of the 570 employees of Grand Tire Service, $\frac{7}{30}$ have given to the United Fund. How many have given to the United Fund?

5. _____

**6.**  A school gives scholarships to $\frac{3}{25}$ of its 1900 freshmen. How many students receive scholarships?

6. _____

**7.**  Kim earns $1500 a month. If she uses $\frac{1}{3}$ of her income on housing, how much does she pay for housing?

7. _____

**8.**  Akiko's home is $\frac{3}{5}$ of the way from Carolyn's home to Laplace College, a distance of 45 miles. How far is it from Carolyn's home to Akiko's?

8. _____

**9.**  Deepak puts $\frac{1}{12}$ of his weekly earnings in a retirement fund. If he makes $1248 a week, how much does he put in his retirement fund each week?

9. _____

**10.**  The sophomore class at Lincoln High School has 312 students. If $\frac{7}{13}$ of the students are boys, how many boys are in the sophomore class?

10. _____

**11.** The Donut Shack sells donuts, bagels, and muffins. During a typical week, they sell 1120 items, of which $\frac{2}{7}$ are muffins. How many muffins does the Donut Shack sell in a typical week?

**11.** _____

**12.** During the month of February $\frac{5}{7}$ of the days had temperatures that were below normal. How many days had below normal temperatures? (Assume that this is not a leap year)

**12.** _____

**13.** Marcie is reading a 360 page book. How many pages has she read if she had completed $\frac{4}{9}$ of the book?

**13.** _____

**14.** A local hospital is recruiting new blood donors. During the month of June, $\frac{3}{16}$ of the people who donated were first-time donors. If 112 people donated blood in June, how many were first time donors?

**14.** _____

**15.** Major league baseball teams play 162 games during the regular season. If a team wins $\frac{15}{27}$ of its games, how many games does it win?

**15.** _____

**16.** A lawnmower uses a gasoline/oil mixture in which $\frac{1}{30}$ of the mixture must be oil. If the tank holds 150 ounces of the mixture, how many ounces are oil?

**16.** _____

**17.** During a local election, a candidate received $\frac{2}{3}$ of the votes. If 2532 people voted in the election, how many votes did the candidate receive?

**17.** _____

**18.** George takes a train to work everyday. The distance from George's house to the train station is $\frac{1}{8}$ of the total distance of 48 miles to work. What is the distance from his home to the train station?

**18.** _____

**19.** Mary must calculate the area of her home office for tax purposes. What is the area of Mary's home office if it takes up $\frac{2}{15}$ of her entire 2340 square foot home?

**19.** _____

**20.** Darrell is participating in a 150-mile bike ride for charity. There are 10 equally-spaced stops including the finish line. How many miles has Darrell completed after reaching the 7$^{\text{th}}$ stop?

**20.** _____

**21.** Elena noticed that her gas gauge moved from the $\frac{7}{8}$ mark to the $\frac{4}{8}$ mark during a recent trip. If her tank holds 16 gallons, how many gallons did she use during this trip?

**21.** _____

**22.** A local college has 8700 undergraduate students and $\frac{7}{30}$ of these students commute to school. How many students commute to school?

**22.** _____

*The Hu family earned $54,000 last year. Use this fact to solve Problems 23–26.*

23. They paid $\frac{1}{3}$ of their income for taxes. How much did they pay in taxes?

23. _____

24. They spend $\frac{2}{5}$ of their income for rent. How much did they spend on rent?

24. _____

25. They saved $\frac{1}{16}$ of their income. How much did they save?

25. _____

26. They spend $\frac{1}{6}$ of their income on food. How much did they spend on food?

26. _____

*The Highview Condo Association collects $96,000 each year from its members. Use this fact to solve Problems 27–30.*

27. The association spends $\frac{1}{6}$ of this money on landscaping. How much do they spend on landscaping?

27. _____

28. The association spends $\frac{5}{12}$ of this money on routine maintenance. How much do they spend on routine maintenance?

28. _____

29. The association spends $\frac{3}{8}$ on insurance. How much to they spend on insurance?

29. _____

**30.** The association puts $\frac{3}{32}$ of this money in an emergency fund. How much money do they put in this fund?

**30.** _____

# Chapter 2 MULTIPLYING AND DIVIDING FRACTIONS

### 2.7    Dividing Fractions

| Learning Objectives |
| --- |
| 1    Find the reciprocal of a fraction. |
| 2    Divide fractions. |
| 3    Solve application problems in which fractions are divided. |

### Key Terms

Use the vocabulary terms listed below to complete each statement in exercises 1–3.

**reciprocals          indicator words       quotient**

1.  Two numbers are _____ of each other if their product is 1.

2.  The words "per" and "divided equally" are _____ for division.

3.  In the problem $192 \div 12 = 16$, 16 is called the _____.

### Objective 1    Find the reciprocal of a fraction.

*Find the reciprocal of each fraction.*

1.  $\dfrac{3}{4}$

1. _____

2.  $\dfrac{9}{2}$

2. _____

3.  $\dfrac{1}{3}$

3. _____

4.  $\dfrac{6}{7}$

4. _____

5.  10

5. _____

6.  $\dfrac{15}{4}$

6. _____

## Objective 2    Divide fractions

*Divide. Write the answer in lowest terms. Change the answers to a whole or mixed number where possible.*

**7.**    $\dfrac{4}{5} \div \dfrac{3}{8}$                                                  **7.** _____

**8.**    $\dfrac{28}{5} \div \dfrac{42}{25}$                                       **8.** _____

**9.**    $\dfrac{\frac{7}{10}}{\frac{14}{5}}$                                               **9.** _____

**10.**    $\dfrac{\frac{4}{9}}{\frac{16}{27}}$                                             **10.** _____

**11.**    $9 \div \dfrac{3}{2}$                                               **11.** _____

**12.**    $\dfrac{5}{8} \div 15$                                             **12.** _____

**13.**    $\dfrac{\frac{11}{3}}{5}$                                               **13.** _____

**14.**    $\dfrac{\frac{6}{11}}{18}$                                               **14.** _____

**15.**    $\dfrac{\frac{8}{15}}{\frac{10}{12}}$                                             **15.** _____

**16.**    $4 \div \dfrac{12}{7}$                                             **16.** _____

**17.**  $\dfrac{3}{4} \div \dfrac{27}{8}$                    17. _____

**18.**  $\dfrac{75}{8} \div 16$                    18. _____

**19.**  $16 \div \dfrac{75}{8}$                    19. _____

## Objective 3    Solve application problems in which fractions are divided.

*Solve each application problem.*

**20.**  Abel has a piece of property with an area of $\frac{7}{8}$ acre. He wishes to divide it into four equal parts for his children. How many acres of land will each child get?                    20. _____

**21.**  Amanda wants to make doll dresses to sell at a craft's fair. Each dress needs $\frac{1}{3}$ yard of material. She has 18 yards of material. Find the number of dresses that she can make.                    21. _____

**22.**  It takes $\frac{4}{5}$ pound of salt to fill a large salt shaker. How many salt shakers can be filled with 32 pounds of salt?                    22. _____

**23.**  Lynn has 2 gallons of lemonade. If each of her Brownies gets $\frac{1}{12}$ gallon of lemonade, how many Brownies does she have?                    23. _____

**24.** How many $\frac{1}{9}$-ounce medicine vials can be filled with 7 ounces of medicine?

**24.** _____

**25.** Each guest at a party will eat $\frac{5}{16}$ pound of chips. How many guests can be served with 10 pounds of chips?

**25.** _____

**26.** Samantha uses $\frac{2}{3}$ yard of ribbon to make a bow for each package she wraps at May's Department Store. How many bows can she make if she has 60 yards of ribbon?

**26.** _____

**27.** Bill wishes to make hamburger patties that weight $\frac{5}{12}$ pound. How many hamburger patties can he make with 10 pounds of hamburger?

**27** _____

**28.** Glen has a small pickup truck that will carry $\frac{3}{4}$ cord of firewood. Find the number of trips needed to deliver 30 cords of wood.

**28.** _____

**29.** How many $\frac{5}{4}$-cup glass tumblers can be filled from a 20-cup bowl of punch?

**29.** _____

**30.** Janine wants to make wooden coasters for glasses. How many coasters can she make from a 12-inch round post if each coaster is to be $\frac{3}{8}$-inch tall?

**30.** _____

# Chapter 2 MULTIPLYING AND DIVIDING FRACTIONS

### 2.8    Multiplying and Dividing Mixed Numbers

| **Learning Objectives** | |
|---|---|
| 1 | Estimate the answer and multiply mixed numbers. |
| 2 | Estimate the answer and divide mixed numbers. |
| 3 | Solve application problems with mixed numbers. |

### Key Terms

Use the vocabulary terms listed below to complete each statement in exercises 1–3.

> **mixed number**          **simplify**      **round**

1.  To _____a fraction means to write the fraction in lowest terms.

2.  If the numerator of a fraction is half of the denominator or more, _____ up to the next whole number to estimate the product of a mixed number and a whole number.

3.  $2\frac{7}{11}$ is an example of a _____.

### Objective 1    Estimate the answer and multiply mixed numbers.

*First estimate the answer. Then multiply to find the exact answer. Simplify all answers.*

1.  $5\frac{1}{3} \cdot 2\frac{1}{2}$

   **1.**
   **Estimate**_____

   **Exact** _____

2.  $3\frac{1}{2} \cdot 4\frac{2}{7}$

   **2.**
   **Estimate**_____

   **Exact** _____

3.  $3\frac{1}{2} \cdot 1\frac{3}{7}$

   **3.**
   **Estimate**_____

   **Exact** _____

4.  $4\frac{4}{9} \cdot 2\frac{2}{5}$

   **4.**
   **Estimate**_____

   **Exact** _____

**5.**    $5\frac{2}{3} \cdot 7\frac{1}{8}$

                                          **5.**

                                          **Estimate**_____

                                          **Exact** _____

**6.**    $18 \cdot 2\frac{5}{9}$

                                          **6.**

                                          **Estimate**_____

                                          **Exact** _____

**7.**    $3\frac{2}{5} \cdot 15$

                                          **7.**

                                          **Estimate**_____

                                          **Exact** _____

**8.**    $\frac{5}{6} \cdot 2\frac{1}{2} \cdot 2\frac{2}{5}$

                                          **8.**

                                          **Estimate**_____

                                          **Exact** _____

**9.**    $1\frac{1}{4} \cdot 1\frac{1}{3} \cdot 1\frac{1}{2}$

                                          **9.**

                                          **Estimate**_____

                                          **Exact** _____

**10.**    $9 \cdot 3\frac{1}{4} \cdot 1\frac{3}{13} \cdot 2\frac{2}{3}$

                                          **10.**

                                          **Estimate**_____

                                          **Exact** _____

## Objective 2    Estimate the answer and divide mixed numbers.

*First estimate the answer. Then divide to find the exact answer. Simplify all answers.*

**11.**    $5\frac{5}{6} \div 5\frac{1}{4}$

                                          **11.**

                                          **Estimate**_____

                                          **Exact** _____

**12.**    $4\frac{5}{8} \div 1\frac{1}{4}$

                                          **12.**

                                          **Estimate**_____

                                          **Exact** _____

**13.**    $4\frac{3}{8} \div 3\frac{1}{2}$

                                          **13.**

                                          **Estimate**_____

                                          **Exact** _____

**14.**    $6\frac{1}{4} \div 2\frac{1}{2}$

**14.**
**Estimate**_____

**Exact** _____

**15.**    $14 \div 8\frac{2}{5}$

**15.**
**Estimate**_____

**Exact** _____

**16.**    $7\frac{1}{3} \div 6$

**16.**
**Estimate**_____

**Exact** _____

**17.**    $7\frac{1}{2} \div \frac{2}{3}$

**17.**
**Estimate**_____

**Exact** _____

**18.**    $2\frac{5}{8} \div 1\frac{3}{4}$

**18.**
**Estimate**_____

**Exact** _____

**19.**    $8\frac{3}{4} \div 5$

**19.**
**Estimate**_____

**Exact** _____

**20.**    $16 \div 2\frac{7}{8}$

**20.**
**Estimate**_____

**Exact** _____

**Objective 3    Solve application problems with mixed numbers.**

*First estimate the answer. Then solve each application problem. Simplify all answers.*

**21.**    Maria wants to make 20 dresses to sell at a bazaar. Each dress needs $3\frac{1}{4}$ yards of material. How many yards does she need?

**21.**
**Estimate**_____

**Exact** _____

**22.** Juan worked $38\frac{1}{4}$ hours at $9 per hour. How much did he make?

**22.**
**Estimate** _____

**Exact** _____

**23.** Each home in an area needs $41\frac{1}{3}$ yards of rain gutter. How much rain gutter would be needed for 6 homes?

**23.**
**Estimate** _____

**Exact** _____

**24.** A farmer applies fertilizer to this fields at a rate of $5\frac{5}{6}$ gallons per acre. How many acres can he fertilize with $65\frac{5}{6}$ gallons?

**24.**
**Estimate** _____

**Exact** _____

**25.** Insect spray is mixed using $1\frac{3}{4}$ ounces of a chemical per gallon of water. How many ounces of the chemical are needed to mix with $28\frac{4}{5}$ gallons of water?

**25.**
**Estimate** _____

**Exact** _____

**26.** How many $\frac{3}{4}$-pound peanut cans can be filled with 15 pounds of peanuts?

**26.**
**Estimate** _____

**Exact** _____

**27.** How many dresses can be made from 70 yards of material if each dress requires $4\frac{3}{8}$ yards?

**27.**
**Estimate** _____

**Exact** _____

**28.** Arnette worked $24\frac{1}{2}$ hours and earned $9 per hour. How much did she earn?

**28.**
**Estimate** _____

**Exact** _____

29. Juan has $3\frac{1}{2}$ sticks of margarine. If each stick weighs $\frac{1}{4}$ pound, how much does Juan's margarine weigh?

29.
**Estimate** _____

**Exact** _____

30. A dental office plays taped music constantly. Each tape takes $1\frac{1}{4}$ hours. How many tapes are played during $7\frac{1}{2}$ hours?

30.
**Estimate** _____

**Exact** _____

# Chapter 3 ADDING AND SUBTRACTING FRACTIONS

## 3.1     Adding and Subtracting Like Fractions

**Learning Objectives**
1      Define like and unlike fractions.
2      Add like fractions.
3      Subtract like fractions.

**Key Terms**

Use the vocabulary terms listed below to complete each statement in exercises 1–2.

         **like fractions**         **unlike fractions**

1.    Fractions with different denominators are called _____.

2.    Fractions with the same denominator are called _____.

**Objective 1    Define like and unlike fractions.**

*Write **like** or **unlike** for each set of fractions.*

1.    $\dfrac{3}{5}, \dfrac{4}{10}$                                     1. _____

2.    $\dfrac{9}{7}, \dfrac{2}{7}$                                       2. _____

3.    $\dfrac{2}{5}, \dfrac{3}{5}$                                       3. _____

4.    $\dfrac{2}{3}, \dfrac{3}{2}$                                       4. _____

5.    $\dfrac{2}{15}, \dfrac{3}{15}, \dfrac{1}{5}$                               5. _____

6.    $\dfrac{18}{7}, \dfrac{21}{7}, \dfrac{7}{7}$                               6. _____

## Objective 2  Add like fractions.

*Add and simplify the answer.*

7.  $\dfrac{5}{8}+\dfrac{1}{8}$

7.  _____

8.  $\dfrac{11}{15}+\dfrac{1}{15}$

8.  _____

9.  $\dfrac{7}{8}+\dfrac{5}{8}$

9.  _____

10.  $\dfrac{4}{3}+\dfrac{7}{3}$

10.  _____

11.  $\dfrac{11}{16}+\dfrac{7}{16}$

11.  _____

12.  $\dfrac{1}{6}+\dfrac{5}{6}$

12.  _____

13.  $\dfrac{1}{5}+\dfrac{2}{5}+\dfrac{4}{5}$

13.  _____

14.  $\dfrac{6}{10}+\dfrac{4}{10}+\dfrac{3}{10}$

14.  _____

15.  $\dfrac{67}{81}+\dfrac{29}{81}+\dfrac{12}{81}$

15.  _____

*Solve each application problem. Write answers in lowest terms.*

16.  Malika walked $\frac{3}{8}$ of a mile downhill and then $\frac{1}{8}$ of a mile along a creek. How far did she walk altogether?

16.  _____

**17.** Last month the Yee family paid $\frac{2}{11}$ of a debt. This month they paid an additional $\frac{5}{11}$ of the same debt. What fraction of the debt has been paid?

17. _____

**18.** Brent painted $\frac{1}{6}$ of a house last week and another $\frac{3}{6}$ this week. How much of the house is painted?

18. _____

**Objective 3    Subtract like fractions.**

*Subtract and simplify the answer.*

**19.** $\frac{3}{10} - \frac{1}{10}$

19. _____

**20.** $\frac{11}{16} - \frac{3}{16}$

20. _____

**21.** $\frac{16}{21} - \frac{2}{21}$

21. _____

**22.** $\frac{16}{15} - \frac{6}{15}$

22. _____

**23.** $\frac{25}{28} - \frac{15}{28}$

23. _____

**24.** $\frac{31}{36} - \frac{11}{36}$

24. _____

**25.** $\frac{91}{100} - \frac{41}{100}$

25. _____

**26.** $\dfrac{736}{400} - \dfrac{496}{400}$                 **26.** _____

**27.** $\dfrac{365}{224} - \dfrac{269}{224}$                 **27.** _____

*Solve each application problem. Write answers in lowest terms.*

**28.** Bill must walk $\frac{9}{12}$ of a mile. He has already walked     **28.** _____
$\frac{1}{12}$ of a mile. How much farther must he walk?

**29.** Jeff planted $\frac{11}{18}$ of his garden in corn and potatoes. If     **29.** _____
$\frac{5}{18}$ of the garden is corn, how much of the garden is
potatoes?

**30.** The Thompsons owe $\frac{8}{15}$ of a debt. If they pay $\frac{2}{15}$ of     **30.** _____
it this month, what fraction of the debt will they still
owe?

# Chapter 3 ADDING AND SUBTRACTING FRACTIONS

### 3.2    Least Common Multiples

| **Learning Objectives** |
| --- |
| 1    Find the least common multiple. |
| 2    Find the least common multiple using multiples of the largest number. |
| 3    Find the least common multiple using prime factorization. |
| 4    Find the least common multiple using an alternative method. |
| 5    Write a fraction with an indicated denominator. |

### Key Terms

Use the vocabulary terms listed below to complete each statement in exercises 1–2.

**least common multiple        LCM**

1.    The _____ of two whole numbers is the smallest whole number divisible by both of the numbers.

2.    _____ is the abbreviation for least common multiple.

### Objective 1    Find the least common multiple.

*Find the least common multiple for each of the following by listing multiples of each number.*

1.    7, 14                                      1. _____

2.    12, 18                                     2. _____

3.    21, 28                                     3. _____

4.    30, 75                                     4. _____

5.    15, 21                                     5. _____

## Objective 2    Find the least common multiple using multiples of the largest number.

*Find the least common multiple for each of the following by using multiples of the largest number.*

**6.**   5, 12                           **6.** _____

**7.**   16, 20                        **7.** _____

**8.**   15, 25                        **8.** _____

**9.**   14, 35                        **9.** _____

**10.**   32, 40                       **10.** _____

## Objective 3    Find the least common multiple using prime factorization.

*Find the least common multiple for each of the following using prime factorization.*

**11.**   14, 48                        **11.** _____

**12.**   28, 32                        **12.** _____

**13.**   10, 24, 32                   **13.** _____

**14.**   16, 20, 25                   **14.** _____

**15.**   7, 12, 21, 35                                **15.** _____

**Objective 4    Find the least common multiple using an alternative method.**

*Find the least common multiple for each of the following using an alternative method.*

**16.**   10, 15                                      **16.** _____

**17.**   22, 55                                      **17.** _____

**18.**   35, 85                                      **18.** _____

**19.**   4, 18, 27                                   **19.** _____

**20.**   12, 30, 40                                  **20.** _____

**21.**   10, 12, 36                                  **21.** _____

**Objective 5    Write a fraction with an indicated denominator.**

*Rewrite each fraction with the indicated denominator.*

**22.**   $\dfrac{1}{9} = \dfrac{}{36}$                **22.** _____

**23.**   $\dfrac{2}{7} = \dfrac{}{63}$

**24.**   $\dfrac{1}{13} = \dfrac{}{78}$

**25.**   $\dfrac{5}{6} = \dfrac{}{72}$

**26.**   $\dfrac{3}{13} = \dfrac{}{52}$

**27.**   $\dfrac{21}{11} = \dfrac{}{55}$

**28.**   $\dfrac{15}{7} = \dfrac{}{84}$

**29.**   $\dfrac{9}{17} = \dfrac{}{102}$

**30.**   $\dfrac{7}{12} = \dfrac{}{60}$

**23.** _____

**24.** _____

**25.** _____

**26.** _____

**27.** _____

**28.** _____

**29.** _____

**30.** _____

# Chapter 3 ADDING AND SUBTRACTING FRACTIONS

### 3.3    Adding and Subtracting Unlike Fractions

| **Learning Objectives** |
| --- |
| 1    Add unlike fractions. |
| 2    Add unlike fractions vertically. |
| 3    Subtract unlike fractions. |
| 4    Subtract unlike fractions vertically. |

## Key Terms

Use the vocabulary terms listed below to complete each statement in exercises 1–2.

**least common denominator**         **LCD**

1.   In order to add or subtract fractions with different denominators, first find the

_____.

2.   _____ is the abbreviation for least common denominator.

## Objective 1    Add unlike fractions.

*Add the following fractions. Simplify all answers.*

1.   $\dfrac{1}{3} + \dfrac{1}{2}$                              1. _____

2.   $\dfrac{1}{5} + \dfrac{5}{8}$                              2. _____

3.   $\dfrac{3}{10} + \dfrac{7}{15}$                             3. _____

4.   $\dfrac{1}{6} + \dfrac{3}{14}$                             4. _____

5.   $\dfrac{4}{15} + \dfrac{9}{20}$                            5. _____

**6.** $\dfrac{1}{4}+\dfrac{2}{7}+\dfrac{3}{14}$

6. _____

**7.** $\dfrac{1}{3}+\dfrac{1}{8}+\dfrac{5}{12}$

7. _____

*Solve each application problem. Write answers in lowest terms.*

**8.** Michael Pippen paid $\dfrac{1}{9}$ of a debt in January, $\dfrac{1}{2}$ in February, $\dfrac{1}{4}$ in March, and $\dfrac{1}{12}$ in April. What fraction of the debt was paid in these four months?

8. _____

**9.** A buyer for a grain company bought $\dfrac{3}{8}$ ton of wheat, $\dfrac{1}{6}$ ton of rice, and $\dfrac{1}{4}$ ton of barley. How many tons of grain were bought?

9. _____

**10.** Find the perimeter (distance around) the figure.

10. _____

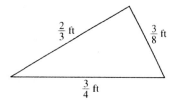

$\frac{2}{3}$ ft     $\frac{3}{8}$ ft

$\frac{3}{4}$ ft

**Objective 2    Add unlike fractions vertically.**

*Add the following fractions. Simplify all answers.*

**11.**    $\dfrac{1}{15}$

$+\dfrac{2}{3}$

11. _____

**12.**  $\dfrac{7}{12}$

$+\dfrac{3}{8}$

**12.** _____

**13.**  $\dfrac{2}{15}$

$+\dfrac{7}{10}$

**13.** _____

**14.**  $\dfrac{1}{6}$

$+\dfrac{2}{9}$

**14.** _____

**15.**  $\dfrac{3}{7}$

$+\dfrac{4}{21}$

**15.** _____

**16.**  $\dfrac{5}{22}$

$+\dfrac{7}{33}$

**16.** _____

**17.**  $\dfrac{6}{13}$

$+\dfrac{15}{52}$

**17.** _____

**18.** $\dfrac{3}{14}$

$+\dfrac{5}{21}$

**18.** _____

**19.** $\dfrac{5}{18}$

$+\dfrac{7}{27}$

**19.** _____

**20.** $\dfrac{1}{12}$

$+\dfrac{1}{8}$

**20.** _____

**Objective 3    Subtract unlike fractions.**
**Objective 4    Subtract unlike fractions vertically.**

*Subtract the following fractions. Simplify all answers.*

**21.** $\dfrac{7}{8}-\dfrac{1}{2}$

**21.** _____

**22.** $\dfrac{5}{8}-\dfrac{1}{6}$

**22.** _____

**23.** $\dfrac{5}{9}$

$-\dfrac{5}{12}$

**23.** _____

**24.** $\dfrac{9}{16}$
  $-\dfrac{3}{10}$

**24.** _____

**25.** $\dfrac{7}{8}$
  $-\dfrac{7}{28}$

**25.** _____

**26.** $\dfrac{3}{5} - \dfrac{1}{4}$

**26.** _____

**27.** $\dfrac{9}{10} - \dfrac{4}{25}$

**27.** _____

*Solve each application problem. Write answers in lowest terms.*

**28.** A company has $\frac{5}{8}$ acre of land. They sold $\frac{1}{3}$ acre. How much land is left?

**28.** _____

**29.** Greg had $\frac{7}{12}$ of his savings goal to complete at the beginning of the month. During the month he saved another $\frac{1}{8}$ of the goal. How much of the goal is left to save?

**29.** _____

**30.**   A $\frac{3}{4}$-inch nail was hammered through a board and          **30.** _____

$\frac{1}{8}$-inch of the nail stuck out. How thick is the board?

# Chapter 3 ADDING AND SUBTRACTING FRACTIONS

### 3.4    Adding and Subtracting Mixed Numbers

| Learning Objectives |
| --- |
| **1**     Estimate an answer, then add or subtract mixed numbers. |
| **2**     Estimate an answer, then subtract mixed numbers by regrouping. |
| **3**     Add or subtract mixed numbers using an alternate method. |

### Key Terms

Use the vocabulary terms listed below to complete each statement in exercises 1–2.

**regrouping when adding fractions**

**regrouping when subtracting fractions**

1.     _____ is the method used in the subtraction of mixed numbers when the fraction part of the minuend is less than the fraction part of the subtrahend.

2.     _____ is the method used in the addition of mixed numbers when the sum of the fraction is greater than 1.

### Objective 1    Estimate an answer, then add or subtract mixed numbers.

*First estimate the answer. Then add or subtract to find the exact answer. Write answers as mixed numbers in lowest terms.*

1.     $5\frac{1}{7}$

       $+\,4\frac{3}{7}$

1.

**Estimate**_____

**Exact** _____

2.     $3\frac{1}{9}$

       $+\,4\frac{7}{8}$

2.

**Estimate**_____

**Exact** _____

3.     $17\frac{5}{8}$

       $12\frac{1}{4}$

       $+\;\;5\frac{5}{6}$

3.

**Estimate**_____

**Exact** _____

**4.**    $126\frac{4}{5}$

$28\frac{9}{10}$

$+\,13\frac{2}{15}$

**4.**

**Estimate** _____

**Exact** _____

**5.**    $26\frac{11}{14}$

$-\,13\frac{5}{18}$

**5.**

**Estimate** _____

**Exact** _____

**6.**    $14\frac{4}{7}$

$-\,8\frac{1}{8}$

**6.**

**Estimate** _____

**Exact** _____

*First estimate the answer. Then solve each application problem.*

**7.**    A painter used $2\frac{1}{3}$ cans of paint one day and $1\frac{7}{8}$ cans the next day. How many cans did he use altogether?

**7.**

**Estimate** _____

**Exact** _____

**8.**    The Eastside Wholesale Vegetable Market sold $4\frac{3}{4}$ tons of broccoli, $8\frac{2}{3}$ tons of spinach, $2\frac{1}{2}$ tons of corn, and $1\frac{5}{12}$ tons of turnips last month. Find the total number of tons of these vegetables sold by the market last month.

**8.**

**Estimate** _____

**Exact** _____

**9.**    Paul worked $12\frac{3}{4}$ hours over the weekend. He worked $6\frac{3}{8}$ hours on Saturday. How many hours did he work on Sunday?

**9.**

**Estimate** _____

**Exact** _____

**10.** On Monday, $7\frac{3}{4}$ tons of cans were recycled, while $9\frac{4}{5}$ tons were recycled on Tuesday. How many more tons were recycled on Tuesday than on Monday?

**10.**
**Estimate** _____

**Exact** _____

**Objective 2    Estimate an answer, then subtract mixed numbers by regrouping.**

*First estimate the answer. Then subtract to find the exact answer. Simplify all answers*

**11.**     $11\frac{1}{4}$
         $-\ 6\frac{3}{4}$

**11.**
**Estimate** _____

**Exact** _____

**12.**     $6\frac{1}{3}$
         $-5\frac{7}{12}$

**12.**
**Estimate** _____

**Exact** _____

**13.**     $12\frac{5}{12}$
         $-11\frac{11}{16}$

**13.**
**Estimate** _____

**Exact** _____

**14.**     $129\frac{2}{3}$
         $-98\frac{14}{15}$

**14.**
**Estimate** _____

**Exact** _____

**15.**     $42$
         $-19\frac{3}{4}$

**15.**
**Estimate** _____

**Exact** _____

**16.**      21

$-17\frac{9}{16}$

**16.**

**Estimate**_____

**Exact** _____


**17.**      $372\frac{5}{6}$

$-208\frac{3}{8}$

**17.**

**Estimate**_____

**Exact** _____


**18.**      147

$-39\frac{5}{6}$

**18.**

**Estimate**_____

**Exact** _____


*First estimate the answer. Then solve each application problem.*

**19.**     Amy Atwood worked 40 hours during a certain week. She worked $8\frac{1}{4}$ hours on Monday, $6\frac{3}{8}$ hours on Tuesday, $7\frac{3}{4}$ hours on Wednesday, and $8\frac{3}{4}$ hours on Thursday. How many hours did she work on Friday?

**19.**

**Estimate**_____

**Exact** _____


**20.**     Three sides of a parking lot are $35\frac{1}{4}$ yards, $42\frac{7}{8}$ yards, and $32\frac{3}{4}$ yards. If the total distance around the lot is $145\frac{1}{2}$ yards, find the length of the fourth side.

**20.**

**Estimate**_____

**Exact** _____

21. A concrete truck is loaded with $11\frac{5}{8}$ cubic yards of concrete. The driver unloads $1\frac{1}{6}$ cubic yards at the first stop, and $2\frac{5}{12}$ cubic yards at the second stop. The customer at the third stop gets 3 cubic yards. How much concrete is left in the truck?

**21.**
**Estimate**_____

**Exact** _____

22. Debbie Andersen bought 15 yards of material at a sale. She made a shirt with $3\frac{1}{8}$ yards of the material, a dress with $4\frac{7}{8}$ yards, and a jacket with $3\frac{3}{4}$ yards. How many yards of material were left over?

**22.**
**Estimate**_____

**Exact** _____

*Find **x** in each figure.*

23. 

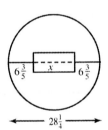

**23.**
**Estimate**_____

**Exact** _____

24. 

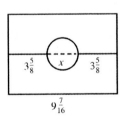

**24.**
**Estimate**_____

**Exact** _____

## Objective 3 Add or subtract mixed numbers using an alternate method.

*Add or subtract by changing mixed numbers to improper fractions. Simplify all answers.*

**25.**  $5\frac{1}{3}$

    $+\,2\frac{5}{6}$

**25.**

**Estimate** _____

**Exact** _____

**26.**  $1\frac{3}{8}$

    $+\,2\frac{3}{5}$

**26.**

**Estimate** _____

**Exact** _____

**27.**  $3\frac{7}{8}$

    $+\,1\frac{5}{12}$

**27.**

**Estimate** _____

**Exact** _____

**28.**  $3\frac{1}{2}$

    $-\,1\frac{2}{3}$

**28.**

**Estimate** _____

**Exact** _____

**29.**  $3\frac{2}{3}$

    $-\,1\frac{5}{6}$

**29.**

**Estimate** _____

**Exact** _____

**30.**  $9\frac{1}{8}$

    $-\,7\frac{4}{9}$

**30.**

**Estimate** _____

**Exact** _____

# Chapter 3 ADDING AND SUBTRACTING FRACTIONS

## 3.5    Order Relations and the Order of Operations

| Learning Objectives | |
|---|---|
| 1 | Identify the greater of two fractions. |
| 2 | Use exponents with fractions. |
| 3 | Use the order of operations with fractions. |

## Key Terms

Use the vocabulary terms listed below to complete each statement in exercises 1–2.

>        <

1.    The symbol _____ means " is less than."

2.    The symbol _____ means " is greater than."

## Objective 1    Identify the greater of two factions.

*Write > or < to make a true statement.*

1.    $\dfrac{1}{2} \underline{\phantom{xx}} \dfrac{5}{8}$

2.    $\dfrac{3}{8} \underline{\phantom{xx}} \dfrac{5}{16}$

3.    $\dfrac{7}{5} \underline{\phantom{xx}} \dfrac{19}{15}$

4.    $\dfrac{5}{12} \underline{\phantom{xx}} \dfrac{3}{5}$

5.    $\dfrac{11}{15} \underline{\phantom{xx}} \dfrac{13}{20}$

6.    $\dfrac{13}{24} \underline{\phantom{xx}} \dfrac{23}{36}$

7.    $\dfrac{23}{40} \underline{\phantom{xx}} \dfrac{17}{30}$

8.    $\dfrac{17}{25} \underline{\phantom{xx}} \dfrac{9}{16}$

1. _____

2. _____

3. _____

4. _____

5. _____

6. _____

7. _____

8. _____

**9.**  $\dfrac{7}{9} \underline{\hspace{1cm}} \dfrac{8}{11}$

9. _____

## Objective 2   Use exponents with fractions.

*Simplify. Write the answer in lowest terms.*

**10.**  $\left(\dfrac{1}{4}\right)^3$

10. _____

**11.**  $\left(\dfrac{1}{2}\right)^2$

11. _____

**12.**  $\left(\dfrac{2}{3}\right)^2$

12. _____

**13.**  $\left(\dfrac{5}{3}\right)^3$

13. _____

**14.**  $\left(\dfrac{3}{2}\right)^4$

14. _____

**15.**  $\left(\dfrac{8}{11}\right)^2$

15. _____

**16.**  $\left(\dfrac{1}{2}\right)^4$

16. _____

**17.**  $\left(\dfrac{12}{7}\right)^2$

17. _____

**18.**  $\left(\dfrac{4}{3}\right)^5$

18. _____

## Objective 3    Use the order of operations with fractions.

*Simplify. Write the answer in lowest terms.*

19.    $\left(\dfrac{2}{3}\right)^2 \cdot 6$             19. _____

20.    $\left(\dfrac{4}{5}\right)^2 \cdot \dfrac{5}{12}$         20. _____

21.    $\left(\dfrac{3}{5}\right)^2 \cdot \left(\dfrac{2}{3}\right)^2$       21. _____

22.    $7 \cdot \left(\dfrac{2}{7}\right)^2 \cdot \left(\dfrac{1}{4}\right)^2$       22. _____

23.    $\dfrac{4}{3} - \dfrac{1}{2} + \dfrac{7}{12}$         23. _____

24.    $\dfrac{1}{2} \cdot \dfrac{4}{5} + \dfrac{2}{3} \cdot \dfrac{9}{5}$      24. _____

25.    $\dfrac{5}{8} - \dfrac{2}{3} \cdot \dfrac{3}{4}$         25. _____

**26.**   $\dfrac{3}{4} \cdot \left( \dfrac{4}{5} + \dfrac{3}{10} \right)$           **26.** _____

**27.**   $\left( \dfrac{8}{7} - \dfrac{9}{14} \right) \div \dfrac{3}{7}$           **27.** _____

**28.**   $\dfrac{8}{7} - \dfrac{9}{14} \div \dfrac{9}{7}$           **28.** _____

**29.**   $\left( \dfrac{4}{7} \right)^2 \cdot \left( \dfrac{3}{2} - \dfrac{5}{8} \right) - \dfrac{1}{21} \cdot \dfrac{3}{4}$      **29.** _____

**30.**   $\left( \dfrac{3}{5} \right)^2 + \dfrac{1}{3} \cdot \left( \dfrac{2}{9} - \dfrac{1}{5} \right) \div \dfrac{1}{15}$     **30.** _____

Name:                                    Date:
Instructor:                              Section:

## Chapter 4 DECIMALS

### 4.1     Reading and Writing Decimals

| Learning Objectives |
|---|
| 1     Write parts of a whole using decimals. |
| 2     Identify the place value of a digit. |
| 3     Read and write decimals in words. |
| 4     Write decimals as fractions or mixed numbers. |

### Key Terms

Use the vocabulary terms listed below to complete each statement in exercises 1–3.

**decimals**          **decimal point**          **place value**

1.   We use _____ to show parts of a whole.

2.   A _____ is assigned to each place to the left or right of
     the decimal point.

3.   The dot that separates the whole number part from the fractional part of a decimal
     number is called the _____.

### Objective 1     Write parts of a whole using decimals.

*Write the portion of each square that is shaded as a fraction, as a decimal, and in words.*

1.

1. _____

_____

_____

2.

2. _____

_____

_____

3.

3. _____

_____

_____

## Objective 2    Identify the place value of a digit.

*Identify the digit that has the given place value.*

**4.**    43.507        tenths

                    hundredths

**5.**    0.42583      hundredths

                    thousandths

**6.**    2.83714      thousandths

                    ten-thousandths

**7.**    302.9651    hundreds

                    hundredths

| | |
|---|---|
| **4.** | _____ |
| | _____ |
| **5.** | _____ |
| | _____ |
| **6.** | _____ |
| | _____ |
| **7.** | _____ |
| | _____ |

*Identify the place value of each digit in these decimals.*

**8.**    0.73        7

                    3

**9.**    37.082      3

                    7

                    0

                    8

                    2

| | |
|---|---|
| **8.** | _____ |
| | _____ |
| **9.** | _____ |
| | _____ |
| | _____ |
| | _____ |
| | _____ |

## Objective 3    Read and write decimals in words.

*Tell how to read each decimal in words.*

**10.**    0.08

**11.**    0.007

**12.**    4.06

| | |
|---|---|
| **10.** | _____ |
| **11.** | _____ |
| **12.** | _____ |

**13.** 3.0014

**13.** _____

**14.** 0.0561

**14.** _____

**15.** 10.835

**15.** _____

**16.** 2.304

**16.** _____

**17.** 97.008

**17.** _____

*Write each decimal in numbers*

**18.** Five and four hundredths

**18.** _____

**19.** Eleven and nine thousandths

**19.** _____

**20.** Thirty eight and fifty-two hundred thousandths

**20.** _____

**21.** Three hundred and twenty-three ten-thousandths

**21.** _____

*Use the table below for exercises 22 and 23.*

| Part Number | Size in Centimeters |
|-------------|---------------------|
| 7-A         | 1.08                |
| 7-B         | 1.58                |
| 7-C         | 0.8                 |
| 8-A         | 7.02                |
| 8-B         | 7.202               |

**22.** Which part number is one and eight hundredths cm?

**22.** _____

**23.** Write in words the size of part number 8-B.

**23.** _____

**Objective 4    Write decimals as fractions or mixed numbers.**

*Write each decimal as a fraction or mixed number in lowest terms.*

24.    0.8                                    24. _____

25.    0.001                                  25. _____

26.    3.6                                    26. _____

27.    20.0005                                27. _____

28.    4.26                                   28. _____

29.    0.95                                   29. _____

30.    80.166                                 30. _____

# Chapter 4 DECIMALS

### 4.2    Rounding Decimals

| Learning Objectives |
| --- |
| 1    Learn the rules for rounding decimals. |
| 2    Round decimals to any given place. |
| 3    Round money amounts to the nearest cent or nearest dollar. |

### Key Terms

Use the vocabulary terms listed below to complete each statement in exercises 1–2.

**rounding**           **decimal places**

1.    _____ are the number of digits to the right of the decimal point.

2.    When we "cut off" a number after a certain place value, we are _____ that number.

### Objective 1    Learn the rules for rounding decimals.

*Select the phrase that makes the sentence correct.*

1.    When rounding a number to the nearest tenth, if the        1. _____
      digit in the hundredths place is 5 or more, round the
      digit in the tenths place (up/down).

2.    When rounding a number to the nearest hundredth,          2. _____
      look at the digit in the (tenth/thousandth) place.

### Objective 2    Round decimals to any given place.

*Round each number to the place indicated.*

3.    17.8937 to the nearest tenth                             3. _____

4.    489.84 to the nearest tenth                              4. _____

5.    785.4982 to the nearest thousandth                       5. _____

6.    43.51499 to the nearest ten-thousandth                   6. _____

7.    54.4029 to the nearest hundredth                         7. _____

**8.**    75.399 to the nearest hundredth                    8. _____

**9.**    989.98982 to the nearest thousandth               9. _____

**10.**   486.496 to the nearest one                         10. _____

*Round to the nearest hundredth and then to the nearest tenth. Remember to always round the original number.*

**11.**   283.0491                                           11. _____

                                                               _____

**12.**   89.525                                             12. _____

                                                               _____

**13.**   21.769                                             13. _____

                                                               _____

**14.**   0.8948                                             14. _____

                                                               _____

**15.**   1.437                                              15. _____

                                                               _____

**16.**   0.0986                                             16. _____

                                                               _____

**17.**   78.695                                             17. _____

                                                               _____

**18.**   108.073                                            18. _____

                                                               _____

**Objective 3    Round money amounts to the nearest cent or nearest dollar.**

*Round to the nearest dollar.*

**19.**   $79.12                                             19. _____

**20.** $28.39                                    **20.** _____

**21.** $225.98                                   **21.** _____

**22.** $4797.50                                  **22.** _____

**23.** $11,839.73                                **23.** _____

**24.** $27,869.57                                **24.** _____

*Round to the nearest cent.*

**25.** $1.2499                                   **25.** _____

**26.** $1.0924                                   **26.** _____

**27.** $112.0089                                 **27.** _____

**28.** $134.20506                                **28.** _____

**29.** $1028.6666                                **29.** _____

**30.** $2096.0149                                **30.** _____

# Chapter 4 DECIMALS

### 4.3    Adding and Subtracting Decimals

| **Learning Objectives** |
| --- |
| **1**    Add decimals. |
| **2**    Subtract decimals. |
| **3**    Estimate the answer when adding or subtracting decimals. |

### Key Terms

Use the vocabulary terms listed below to complete each statement in exercises 1–2.

**estimating            front end rounding**

1.    With _____, we round to the highest possible place.

2.    Avoid common errors in working decimal problems by _____ the answer first.

### Objective 1    Add decimals.

*Find each sum.*

1.    $43.96 + 48.53$                           1. _____

2.    $47.94 + 102.38 + 27.631$               2. _____

3.    $39.87 + 25.2 + 40.36$                  3. _____

4.    $87.6 + 90.4$                           4. _____

5.    $45.83 + 20.923 + 5.7$                  5. _____

6.    $4 + 7.99 + 3.46$                       6. _____

**7.**     $10.82 + 5.9 + 4.7 + 6.3 + 20.63$                    **7.** _____

*Find the perimeter of (distance around) each geometric figure by adding the lengths of the sides.*

**8.** 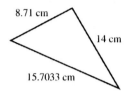                                    **8.** _____

**9.** 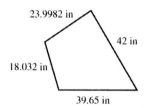                                    **9.** _____

## Objective 2    Subtract decimals.

*Find each difference.*

**10.**     $84.6 - 18.1$                                       **10.** _____

**11.**     $223.3 - 107.5$                                     **11.** _____

**12.**     $41.2 - 8.76$                                       **12.** _____

**13.**     $69.524 - 26.958$                                   **13.** _____

**14.**     $23.104 - 6.98$                                     **14.** _____

**15.**   71 – 12.68

**15.** _____

**16.**   689 – 79.832

**16.** _____

*Find the unknown measurement in each figure.*

**17.**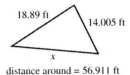
distance around = 56.911 ft

**17.** _____

**18.**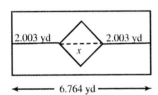

**18.** _____

**Objective 3    Estimate the answer when adding or subtracting decimals.**

*First, use front end rounding and estimate each answer. Then add or subtract to find the exact answer.*

**19.**   32.99
         41.72
       +  8.2

**19.**
**Estimate** _____

**Exact** _____

**20.**   20.85
        –  7.69

**20.**
**Estimate** _____

**Exact** _____

**21.**   9.7
       –  4.862

**21.**
**Estimate** _____

**Exact** _____

**22.**      593.8

     27.93

  + 54.87

**22.**

**Estimate** _____

**Exact** _____

**23.**      9

  − 3.47

**23.**

**Estimate** _____

**Exact** _____

*First, use front end rounding and estimate each answer. Then add or subtract to find the exact answer.*

**24.**  Kim spent $28.25 for books, $29.47 for a blouse, and $17.85 for a compact disk. How much did she spend?

**24.**

**Estimate** _____

**Exact** _____

**25.**  Manuel has agreed to work 27.5 hours at a certain job. He has already worked 9.65 hours. How many hours does he have left to work?

**25.**

**Estimate** _____

**Exact** _____

**26.**  At a fruit stand, Lynn Knight bought $8.53 worth of apples, $11.10 worth of peaches, and $28.29 worth of pears. How much did she spend altogether?

**26.**

**Estimate** _____

**Exact** _____

**27.**  A customer gives a clerk a $20 bill to pay for $11.29 in purchases. How much change should the customer get?

**27.**

**Estimate** _____

**Exact** _____

**28.**  A man receives a bill for $83.26 from Exxon. Of this amount, $53.29 is for a tune-up and the rest is for gas. How much did he pay for gas?

**28.**
**Estimate** _____

**Exact** _____

**29.**  At the beginning of a trip to El Cerrito, a car odometer read 80,447.5 miles. It is 81.9 miles to El Cerrito. What should the odometer read after driving to El Cerrito and back?

**29.**
**Estimate** _____

**Exact** _____

**30.**  In India, there are about 50.6 million Internet users, while in Japan, there are about 86.3 million Internet users. How many fewer Internet users are there in India compared to Japan?

**30.**
**Estimate** _____

**Exact** _____

## Chapter 4 DECIMALS

### 4.4    Multiplying Decimals

| **Learning Objectives** |
| --- |
| 1    Multiply decimals. |
| 2    Estimate the answer when multiplying decimals. |

### Key Terms

Use the vocabulary terms listed below to complete each statement in exercises 1–3.

**decimal places        factor          product**

1.    Each number in a multiplication problem is called a _____.

2.    When multiplying decimal numbers, first find the total number of
       _____ in both factors.

3.    The answer to a multiplication problem is called the _____.

### Objective 1    Multiply decimals.

*Find each product.*

1.    0.053
      × 4.3                                          1. _____

2.    0.682
      × 3.9                                          2. _____

3.    19.3
      × 4.7                                          3. _____

4.    96.5
      × 4.6                                          4. _____

**5.**      67.6
        $\times\, 0.023$

**5.** _____

**6.**      906
        $\times\, 0.081$

**6.** _____

**7.**    (0.074)(0.05)

**7.** _____

**8.**    (0.0009)(0.014)

**8.** _____

*In each of the following, find the amount of money earned on a job by multiplying the number of hours worked and the pay per hour. Round your answer to the nearest cent, if necessary.*

**9.**    27 hours at $6.04 per hour

**9.** _____

**10.**    31.6 hours at $9.83 per hour

**10.** _____

*Find the cost of each of the following.*

**11.**    16 apples at $0.59 each

**11.** _____

**12.**    7 quarts of oil at $1.05 each

**12.** _____

*Use the fact that $86 \times 5 = 430$ to solve exercises 13–18 by simply counting decimal places and writing the decimal point in the correct location.*

**13.**    $86 \times 0.5$

**13.** _____

**14.**    $0.86 \times 5$

**14.** _____

**15.**    $8.6 \times 0.05$                             **15.** _____

**16.**    $0.086 \times 0.05$                      **16.** _____

**17.**    $8.6 \times 0.0005$                     **17.** _____

**18.**    $0.0086 \times 0.005$                  **18.** _____

*Solve. Round to the nearest cent, if necessary.*

**19.**    The width of Jane's garden is 15.4 feet and the length of the garden is 22.6 feet. What is the area of her garden?       **19.** _____

**20.**    Steve's car payment is $309.56 per month for 48 months. How much will he pay altogether?       **20.** _____

**21.**    The Duncan family's state income tax is found by multiplying the family income of $32,906.15 by the decimal 0.064. Find their tax.       **21.** _____

**22.**    A recycling center pays $0.142 per pound of aluminum. How much would be paid for 176.3 pounds?       **22.** _____

**Objective 2**    **Estimate the answer when multiplying decimals.**

*First use front-end rounding and estimate the answer. Then multiply to find the exact answer.*

**23.**    $\begin{array}{r} 49.7 \\ \times\, 5.8 \\ \hline \end{array}$                     **23.**

                                               **Estimate**_____

                                               **Exact** _____

**24.**     29.8
          $\times\, 3.4$

**25.**     58.73
          $\times\, 3.72$

**26.**     32.53
          $\times\, 23.26$

**27.**     76.4
          $\times\, 0.57$

**28.**     2.99
          $\times\, 3.5$

**29.**     391.9
          $\times\, 7.74$

**30.**     27.5
          $\times\, 11.2$

**24.**
**Estimate** _____

**Exact** _____

**25.**
**Estimate** _____

**Exact** _____

**26.**
**Estimate** _____

**Exact** _____

**27.**
**Estimate** _____

**Exact** _____

**28.**
**Estimate** _____

**Exact** _____

**29.**
**Estimate** _____

**Exact** _____

**30.**
**Estimate** _____

**Exact** _____

# Chapter 4 DECIMALS

### 4.5 Dividing Decimals

| **Learning Objectives** |
| --- |
| 1 Divide a decimal by a whole number. |
| 2 Divide a number by a decimal. |
| 3 Estimate the answer when dividing decimals. |
| 4 Use the order of operations with decimals. |

### Key Terms

Use the vocabulary terms listed below to complete each statement in exercises 1–4.

**repeating decimal    quotient    dividend    divisor**

1. In a division problem, the number being divided is called the _____.

2. The number $0.8\overline{3}$ is an example of a _____.

3. The answer to a division problem is called the _____.

4. In the problem $6.39 \div 0.9$, 0.9 is called the _____.

### Objective 1    Divide a decimal by a whole number.

*Find each quotient. Round answers to the nearest thousandth, if necessary.*

1.  $6\overline{)10.763}$

1. _____

2.  $5\overline{)34.8}$

2. _____

3.  $33\overline{)77.847}$

3. _____

**4.** $11\overline{)46.98}$                        **4.** _____

**5.** $54\overline{)895.79}$               **5.** _____

*Solve. Round to the nearest cent, if necessary.*

**6.** To build a barbecue, Diana Jenkins bought 589     **6.** _____
bricks, paying $185.70. Find the cost per brick.

**Objective 2**   **Divide a number by a decimal.**

*Find each quotient. Round answers to the nearest thousandth, if necessary.*

**7.** $0.9\overline{)3.4166}$             **7.** _____

**8.** $3.4\overline{)436.05}$             **8.** _____

**9.** $2859.4 \div 0.053$          **9.** _____

**10.** $0.07 \div 0.00043$        **10.** _____

*Solve each application problem. Round money answers to the nearest cent, if necessary.*

**11.** Leon Williams drove 542.2 miles on the 16.3 gallons of gas in his Ford Taurus. How many miles per gallon did he get? Round to the nearest tenth.

**11.** _____

**12.** Lakesha Starr bought 7.4 yards of fabric, paying a total of $26.27. Find the cost per yard.

**12.** _____

**Objective 3  Estimate the answer when dividing decimals.**

*Decide if each answer is **reasonable** or **unreasonable** by rounding the numbers and estimating the answer.*

**13.** $49.8 \div 7.1 = 7.014$

**13.** _____

**14.** $126.2 \div 11.2 = 11.268$

**14.** _____

**15.** $31.5 \div 8.4 = 37.5$

**15.** _____

**16.** $486.9 \div 5.06 = 962.253$

**16.** _____

**17.** $1092.8 \div 37.92 = 2.882$

**17.** _____

**18.** $1564.9 \div 50.049 = 312.674$

**18.** _____

**19.** $8695.15 \div 98.762 = 88.0415$

**19.** _____

**20.** $6608.04 \div 415.6 = 15.9$

**20.** _____

## Objective 4    Use the order of operations with decimals.

*Use the order of operations to simplify each expression.*

**21.**    $3.7 + 5.1^2 - 9.4$

21. _____

**22.**    $3.1^2 - 1.9 + 5.8$

22. _____

**23.**    $42.92 \div 5.8 \times 7.3$

23. _____

**24.**    $55.744 \div (6.4 \times 1.9)$

24. _____

**25.**    $18.5 + (37.1 - 29.8)(10.7)$

25. _____

**26.**    $58.1 - (17.9 - 15.2) \times 1.8$

26. _____

**27.**    $27.51 - 3.2 \times 9.8 \div 1.6$

27. _____

**28.**    $9.1 - 0.07(2.1 \div 0.042)$

28. _____

**29.**    $9.8 \times 4.76 + 17.94 \div 2.6$

29. _____

**30.**    $62.699 \div 7.42 + 3.6 \times 1.4$

30. _____

## Chapter 4 DECIMALS

### 4.6     Writing Fractions as Decimals

| **Learning Objectives** |
| --- |
| 1     Write fractions as equivalent decimals. |
| 2     Compare the size of fractions and decimals. |

### Key Terms

Use the vocabulary terms listed below to complete each statement in exercises 1–4.

   **numerator     denominator          mixed number          equivalent**

1.     A fraction and a decimal that represent the same portion of a whole are
       _____.

2.     The _____ of a fraction is the dividend.

3.     The _____ of a fraction shows the number of equal parts in a
       whole.

4.     A _____ consists of a whole number part and a fractional or
       decimal part.

### Objective 1     Write fractions as equivalent decimals.

*Write each fraction or mixed number as a decimal. Round to the nearest thousandth, if necessary.*

1.     $6\frac{1}{2}$                                        1. _____

2.     $\frac{1}{5}$                                         2. _____

3.     $2\frac{2}{3}$                                        3. _____

4.     $\frac{1}{8}$                                         4. _____

5.     $\frac{1}{11}$                                        5. _____

6.     $7\frac{1}{10}$                                       6. _____

**7.**   $\dfrac{3}{5}$                                      7. _____

**8.**   $\dfrac{7}{8}$                                      8. _____

**9.**   $4\frac{1}{9}$                                     9. _____

**10.**  $\dfrac{13}{25}$                                  10. _____

**11.**  $\dfrac{3}{20}$                                   11. _____

**12.**  $31\frac{3}{13}$                                   12. _____

**13.**  $19\frac{17}{24}$                                  13. _____

**14.**  Jose got 4 hits in 11 times at bat. Write his batting       14. _____
         average as a decimal rounded to the nearest
         thousandth.

## Objective 2    Compare the size of fractions and decimals.

*Write < or > to make a true statement.*

**15.**  $\dfrac{5}{8}$ ___ 0.634                          15. _____

**16.**  $\dfrac{5}{6}$ ___ 0.83                           16. _____

**17.**  $\dfrac{1}{25}$ ___ 0.039                         17. _____

**18.**  $\dfrac{3}{16}$ ___ 0.188                         18. _____

**19.**  $\dfrac{5}{9}$ ___ 0.55                           19. _____

**20.**  $\dfrac{3}{8}$ ___ 0.38

20. _____

**21.**  $0.\overline{7}$ ___ 0.7

21. _____

**22.**  Candy bar A weighs 1.4 ounces, while candy bar B weighs $1\dfrac{3}{8}$ ounces. Which weighs more?

22. _____

*Arrange in order from smallest to largest.*

**23.**  $\dfrac{7}{15}$, 0.466, $\dfrac{9}{19}$

23. _____

**24.**  $\dfrac{8}{9}$, 0.88, 0.89

24. _____

**25.**  $\dfrac{3}{11}$, $\dfrac{1}{3}$, 0.29

25. _____

**26.**  $\dfrac{1}{7}$, $\dfrac{3}{16}$, 0.187

26. _____

**27.**  0.8462, $\dfrac{11}{13}$, $\dfrac{6}{7}$

27. _____

**28.**  1.085, $1\dfrac{5}{11}$, $1\dfrac{7}{20}$

28. _____

**29.**  0.16666, $\dfrac{1}{6}$, 0.1666, 0.01666

29. _____

**30.**  Five cyclists in a race had the following times:
Catherine, 11.06 min      Edita, 10.53 min
Olga, 11.24 min           Diana, 10.51
Anna, 10.38 min
List them in the order they placed.

30. _____

## Chapter 5 RATIO AND PROPORTION

### 5.1    Ratios

| Learning Objectives |
| --- |
| 1    Write ratios as fractions. |
| 2    Solve ratio problems involving decimals or mixed numbers. |
| 3    Solve ratio problems after converting units. |

### Key Terms

Use the vocabulary terms listed below to complete each statement in exercises 1–2.

     **denominator**         **numerator**         **ratio**

1.   A _____ can be used to compare two measurements with the same type of units.

2.   When writing the ratio to compare the width of a room to its height, the width goes in the _____ and the height goes in the _____.

### Objective 1    Write ratios as fractions.

*Write each ratio as a fraction in lowest terms.*

1.   18 to 24                                          1. _____

2.   76 to 101                                         2. _____

3.   125 cents to 95 cents                            3. _____

4.   80 miles to 30 miles                             4. _____

5.   $85 to $135                                      5. _____

6.   5 men to 20 men                                  6. _____

*Solve. Write each ratio as a fraction in lowest terms.*

7.   Mr. Williams is 42 years old, and his son is 18. Find        7. _____
     the ratio of Mr. Williams' age to his son's age.

**8.** When using Roundup vegetation control, add 128 ounces of water for every 6 ounces of the herbicide. Find the ratio of herbicide to water.

8. _____

## Objective 2    Solve ratio problems involving decimals or mixed numbers.

*Write each ratio as a fraction in lowest terms.*

**9.** $6\frac{1}{2}$ to 2

9. _____

**10.** $4\frac{1}{8}$ to 3

10. _____

**11.** 3 to $2\frac{1}{2}$

11. _____

**12.** 11 to $2\frac{4}{9}$

12. _____

**13.** $1\frac{1}{4}$ to $1\frac{1}{2}$

13. _____

**14.** $3\frac{1}{2}$ to $1\frac{3}{4}$

14. _____

*Solve. Write each ratio as a fraction in lowest terms.*

**15.** One refrigerator holds $3\frac{3}{4}$ cubic feet of food, while another holds 5 cubic feet. Find the ratio of the amount of storage in the first refrigerator to the amount of storage in the second.

15. _____

**16.** One car has a $15\frac{1}{2}$ gallon gas tank while another has a 22 gallon gas tank. Find the ratio of the amount the first tank holds to the amount the second tank holds.

16. _____

**17.** The price of gasoline increased from $2.75 per gallon to $3.25 per gallon. Find the ratio of the increase in price to the original price.

17. _____

**18.** The amount of cereal in a giant-size box decreased from 27.6 ounces to 22.4 ounces. Find the ratio of the original size to the new size.

**18.** _____

*For each triangle, find the ratio of the length of the longest side to the length of the shortest side. Write each ratio as a fraction in lowest terms.*

**19.**

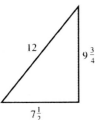

**19.** _____

**20.**

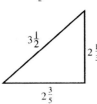

**20.** _____

## Objective 3    Solve ratio problems after converting units.

*Write each ratio as a fraction in lowest terms. Be sure to convert units as necessary.*

**21.** 4 days to 2 weeks

**21.** _____

**22.** 4 feet to 15 inches

**22.** _____

**23.** 6 yards to 10 feet

**23.** _____

**24.** 7 gallons to 8 quarts

**24.** _____

**25.** 40 ounces to 3 pounds

**25.** _____

**26.**    80 cents to \$3

**26.** _____

**27.**    Find the ratio of $17\frac{1}{2}$ inches to $2\frac{1}{3}$ feet.

**27.** _____

**28.**    What is the ratio of $59\frac{1}{2}$ ounces to $4\frac{1}{4}$ pounds?

**28.** _____

**29.**    What is the ratio of $9\frac{1}{3}$ yards to $3\frac{1}{2}$ feet?

**29.** _____

**30.**    A tree is $28\frac{3}{4}$ feet tall. It casts a shadow 81 inches long. Find the ratio of the height of the tree to the length of its shadow.

**30.** _____

# Chapter 5 RATIO AND PROPORTION

## 5.2    Rates

| Learning Objectives |
| --- |
| 1      Write rates as fractions. |
| 2      Find unit rates. |
| 3      Find the best buy based on cost per unit. |

### Key Terms

Use the vocabulary terms listed below to complete each statement in exercises 1–3.

**rate         unit rate       cost per unit**

1. When the denominator of a rate is 1, it is called a _____.

2. The _____ is that rate that tells how much is paid for one item.

3. A _____ compares two measurements with different units.

### Objective 1    Write rates as fractions.

*Write each rate as a fraction in lowest terms.*

1. 75 miles in 25 minutes                     1. _____

2. 85 feet in 17 seconds                      2. _____

3. 28 dresses for 4 women                    3. _____

4. 70 horses for 14 teams                     4. _____

5. 45 gallons in 3 hours                       5. _____

6. 225 miles on 15 gallons                    6. _____

7. 119 pills for 17 patients                   7. _____

8. 144 kilometers on 16 liters                8. _____

**9.**     256 pages for 8 chapters                          9. _____

**10.**    990 miles in 18 hours                             10. _____

## Objective 2    Find unit rates.

*Find each unit rate.*

**11.**    $75 in 5 hours                                    11. _____

**12.**    $3500 in 20 days                                 12. _____

**13.**    $1540 in 14 days                                 13. _____

**14.**    $7875 for 35 pounds                              14. _____

**15.**    $122.76 in 9 hours                               15. _____

**16.**    189.88 miles on 9.4 gallons                      16. _____

*Solve each application problem.*

**17.**    Eric can pack 12 crates of berries in 24 minutes.     17. _____
           Give his rate in crates per minute and in minutes per
           crate.                                                 _____

**18.**    Michelle can plow 7 acres in 14 hours. Give her rate   18. _____
           in acres per hour and in hours per acre.
                                                                  _____

**19.**    Meili earns $220.32 in 24 hours. What is her rate per  19. _____
           hour?

**20.** The 4.6 yards of fabric needed for a dress costs $27.14. Find the cost of 1 yard.

**20.** _____

**21.** A company pays $3225 in dividends for the 1250 shares of its stock. Find the value of dividends per share.

**21.** _____

**22.** Jojo drove 434 miles on 15.5 gallons of gasoline. How many miles did he drive per gallon?

**22.** _____

**23.** It took Marlene $7\frac{1}{4}$ hours to drive 450 miles. What was her average speed per hour?

**23.** _____

**Objective 3    Find the best buy based on cost per unit.**

*Find the best buy (based on cost per unit) for each item.*

**24.** Beans: 12 ounces for $1.49; 16 ounces of $1.89

**24.** _____

**25.** Orange juice: 16 ounces for $0.89; 32 ounces for $1.90

**25.** _____

**26.** Peanut butter: 18 ounces for $1.77; 24 ounces for $2.08

**26.** _____

**27.** Batteries: 4 for $2.79; 10 for $4.19

**27.** _____

**28.** Cola: 6 cans for $1.98; 12 cans for $3.59;
24 cans for $8

**28.** _____

**29.** Soup: 3 cans for $1.75; 5 cans for $2.75;
8 cans for $4.55

**29.** _____

**30.** Cereal: 10 ounces for $1.34; 15 ounces for $1.76;
20 ounces for $2.29

**30.** _____

# Chapter 5 RATIO AND PROPORTION

### 5.3    Proportions

| Learning Objectives |
| --- |
| 1      Write proportions. |
| 2      Determine whether proportions are true or false. |
| 3      Find cross products. |

## Key Terms

Use the vocabulary terms listed below to complete each statement in exercises 1–2.

**cross products          proportion**

1.    A _____ shows that two ratios or rates are
      equivalent.

2.    To see whether a proportion is true, determine if the _____ are
      equal.

## Objective 1    Write proportions.

*Write each proportion.*

1.    11 is to 15 as 22 is to 30.                          1. _____

2.    50 is to 8 as 75 is to 12.                           2. _____

3.    24 is to 30 as 8 is to 10.                           3. _____

4.    36 is to 45 as 8 is to 10.                           4. _____

5.    14 is to 21 as 10 is to 15.                          5. _____

6.    3 is to 33 as 12 is to 132.                          6. _____

**7.** $1\frac{1}{2}$ is to 4 as 21 is to 56.

7. _____

**8.** $3\frac{2}{3}$ is to 11 as 10 is to 30.

8. _____

**9.** $6\frac{2}{5}$ is to 12 as 8 is to 3.

9. _____

**Objective 2   Determine whether proportions are true or false.**

*Determine whether each proportion is true or false by writing the ratios in lowest terms. Show the simplified ratios and then write **true** or **false**.*

**10.** $\dfrac{6}{100} = \dfrac{3}{50}$

10. _____

**11.** $\dfrac{48}{36} = \dfrac{3}{4}$

11. _____

**12.** $\dfrac{3}{8} = \dfrac{21}{28}$

12. _____

**13.** $\dfrac{30}{25} = \dfrac{6}{5}$

13. _____

**14.** $\dfrac{390}{100} = 27$

14. _____

**15.** $\dfrac{35}{21} = \dfrac{3}{4}$

15. _____

**16.** $\dfrac{28}{6} = \dfrac{42}{9}$

16. _____

**17.** $\dfrac{54}{30} = \dfrac{108}{60}$

17. _____

**18.** $\dfrac{15}{24} = \dfrac{25}{35}$

18. _____

**19.** $\dfrac{63}{18} = \dfrac{56}{14}$

19. _____

**20.** $\dfrac{108}{225} = \dfrac{24}{50}$

20. _____

**Objective 3    Find cross products.**

*Use cross products to determine whether each proportion is true or false. Show the cross products and then write **true** or **false**.*

**21.** $\dfrac{10}{45} = \dfrac{6}{27}$

21. _____

**22.** $\dfrac{28}{50} = \dfrac{49}{75}$

22. _____

**23.** $\dfrac{132}{24} = \dfrac{11}{3}$

23. _____

**24.** $\dfrac{3\frac{1}{2}}{4} = \dfrac{14}{16}$

**24.** _____

**25.** $\dfrac{4\frac{3}{5}}{9} = \dfrac{18\frac{2}{5}}{36}$

**25.** _____

**26.** $\dfrac{21}{28} = \dfrac{5\frac{3}{4}}{7}$

**26.** _____

**27.** $\dfrac{22}{54} = \dfrac{6\frac{1}{3}}{5\frac{2}{11}}$

**27.** _____

**28.** $\dfrac{69.9}{3} = \dfrac{100.19}{4.3}$

**28.** _____

**29.** $\dfrac{2.98}{7.1} = \dfrac{1.7}{4.3}$

**29.** _____

**30.** $\dfrac{42.2}{106.8} = \dfrac{84.9}{206}$

**30.** _____

# Chapter 5 RATIO AND PROPORTION

### 5.4    Solving Proportions

| Learning Objectives |
| --- |
| 1    Find the unknown number in a proportion. |
| 2    Find the unknown number in a proportion with mixed numbers or decimals. |

### Key Terms

Use the vocabulary terms listed below to complete each statement in exercises 1–3.

**cross products        proportion    ratio**

1.    A _____ is a statement that two ratios are equal.

2.    The _____ of the proportion $\frac{a}{b} = \frac{c}{d}$ are $ad$ and $bc$.

3.    A _____ is a comparison of two quantities with the same units.

### Objective 1    Find the unknown number in a proportion.

*Find the unknown number in each proportion.*

1.    $\dfrac{3}{2} = \dfrac{x}{6}$                                         1. _____

2.    $\dfrac{9}{4} = \dfrac{36}{x}$                                        2. _____

3.    $\dfrac{9}{7} = \dfrac{x}{28}$                                        3. _____

4.    $\dfrac{x}{11} = \dfrac{44}{121}$                                     4. _____

5.    $\dfrac{35}{x} = \dfrac{5}{3}$                                        5. _____

**6.** $\dfrac{x}{52} = \dfrac{5}{13}$

**6.** _____

**7.** $\dfrac{96}{60} = \dfrac{8}{x}$

**7.** _____

**8.** $\dfrac{7}{5} = \dfrac{98}{x}$

**8.** _____

**9.** $\dfrac{9}{14} = \dfrac{x}{70}$

**9.** _____

**10.** $\dfrac{90}{x} = \dfrac{15}{8}$

**10.** _____

**11.** $\dfrac{x}{110} = \dfrac{7}{10}$

**11.** _____

**12.** $\dfrac{14}{x} = \dfrac{21}{18}$

**12.** _____

**13.** $\dfrac{18}{81} = \dfrac{4}{x}$

**13.** _____

**14.** $\dfrac{100}{x} = \dfrac{75}{30}$

**14.** _____

**15.**   $\dfrac{x}{45} = \dfrac{132}{180}$

**15.** _____

**Objective 2**   **Find the unknown number in a proportion with mixed numbers or decimals.**

*Find the unknown number in each proportion. Write answers as a whole or a mixed number if possible.*

**16.**   $\dfrac{2}{3\frac{1}{4}} = \dfrac{8}{x}$

**16.** _____

**17.**   $\dfrac{3}{x} = \dfrac{5}{1\frac{2}{3}}$

**17.** _____

**18.**   $\dfrac{x}{6} = \dfrac{5\frac{1}{4}}{7}$

**18.** _____

**19.**   $\dfrac{1\frac{1}{5}}{\frac{1}{2}} = \dfrac{6}{x}$

**19.** _____

**20.**   $\dfrac{0}{5\frac{1}{3}} = \dfrac{x}{5}$

**20.** _____

**21.**   $\dfrac{x}{7\frac{1}{2}} = \dfrac{3}{6\frac{2}{3}}$

**21.** _____

**22.**   $\dfrac{3}{x} = \dfrac{0.8}{5.6}$

**22.** _____

**23.** $\dfrac{16}{12} = \dfrac{2}{x}$

**24.** $\dfrac{4.2}{x} = \dfrac{0.6}{2}$

**25.** $\dfrac{2\frac{1}{2}}{1\frac{2}{3}} = \dfrac{x}{2}$

**26.** $\dfrac{2\frac{5}{9}}{x} = \dfrac{23}{\frac{3}{5}}$

**27.** $\dfrac{10}{x} = \dfrac{2\frac{1}{2}}{2}$

**28.** $\dfrac{x}{7.9} = \dfrac{0}{47.4}$

**29.** $\dfrac{x}{4.8} = \dfrac{1.5}{1.2}$

**30.** $\dfrac{32}{2.4} = \dfrac{x}{3}$

## Chapter 5 RATIO AND PROPORTION

## 5.5     Solving Application Problems with Proportions

**Learning Objectives**
1        Use proportions to solve application problems.

**Key Terms**

Use the vocabulary terms listed below to complete each statement in exercises 1–2.

       **rate**            **ratio**

1.    A statement that compares a number of inches to a number of inches is a

       _____ .

2.    A statement that compares a number of gallons to a number of miles is a

       _____ .

**Objective 1    Use proportions to solve application problems.**

*Set up and solve a proportion for each problem.*

1.    A gardening service charges $45 to install 50 square     **1.** _____
feet of sod. Find the charge to install 125 feet.

2.    On a road map, a length of 3 inches represents a      **2.** _____
distance of 8 miles. How many inches represent a
distance of 32 miles?

3.    If 6 melons cost $9, find the cost of 10 melons.       **3.** _____

4.    If 22 hats cost $198, find the cost of 12 hats.       **4.** _____

**5.**    6 pounds of grass seed cover 4200 square feet of ground. How many pounds are needed for 5600 square feet.

5. _____

**6.**    Margie earns $168.48 in 26 hours. How much does she earn in 40 hours?

6. _____

**7.**    Juan makes $477.40 in 35 hours. How much does he make in 60 hours?

7. _____

**8.**    If 5 ounces of a medicine must be mixed with 12 ounces of water, how many ounces of medicine would be mixed with 132 ounces of water?

8. _____

**9.**    The distance between two cities on a road map is 5 inches. The two cities are really 600 miles apart. The distance between two other cities on the map is 8 inches. How many miles apart are these cities?

9. _____

**10.**    The distance between two cities is 600 miles. On a map the cities are 10 inches apart. Two other cities are 720 miles apart. How many inches apart are they on the map?

10. _____

**11.** If 2 visits to a salon cost $80, find the cost of 11 visits.

**11.** _____

**12.** If a 4-minute phone call costs $0.96, find the cost of a 10-minute call.

**12.** _____

**13.** If 150 square yards of carpet cost $3142.50, find the cost of 210 square yards of the carpet.

**13.** _____

**14.** Scott paid $240,000 for a 5-unit apartment house. Find the cost of a 16-unit apartment house.

**14.** _____

**15.** Brian plants his seeds early in the year. To keep them from freezing, he covers the ground with black plastic. A piece with an area of 80 square feet costs $14. Find the cost of a piece with an area of 700 square feet.

**15.** _____

**16.** A taxi ride of 7 miles costs $9.45. Find the cost of a ride of 12 miles.

**16.** _____

**17.** Dog food for 8 dogs costs $15. Find the cost of dog food for 12 dogs.

**17.** _____

**18.** To make battery acid, Jeff mixes $9\frac{1}{2}$ gallons of pure acid with 25 gallons of water. How much acid would be needed for 75 gallons of water?

**18.** _____

**19.** Tax on an $18,000 car is $1620. Find the tax on a $24,000 car.

**19.** _____

**20.** If $18\frac{3}{4}$ yards of material are needed for 5 dresses, how much material is needed for 9 dresses?

**20.** _____

**21.** If it takes 6 minutes to read 4 pages of a book, how long will it take to read 320 pages?

**21.** _____

**22.** If a gallon of paint will cover 400 square feet, how many gallons are needed to cover 2200 square feet?

**22.** _____

**23.** The height of the water in a fish tank rises at a constant rate of 2 inches every 5 minutes. How many minutes will it take to fill the tank if the height must reach 25 inches?

**23.** _____

**24.** It costs $15 dollars to park for 4 hours. How long will you have parked a car if your cost is $25 dollars?

**24.** _____

**25.** A ball that is dropped from a height of 60 inches will rebound to a height of 48 inches. How high will a ball rebound that is dropped from a height of 96 inches?

**25.** _____

**26.** A person weighing 150 pounds on Earth weighs approximately 25 pounds on the moon. How much will a person weigh on Earth if their moon weight is 32 pounds?

**26.** _____

**27.** Five apples cost $1.60. How much will 8 apples cost?

**27.** _____

**28.** A biologist tags 50 deer and releases them in a wildlife preserve area. Over the course of a two-week period, she observes 80 deer, of which 12 are tagged. What is the estimate for the population of deer in this particular area?

**28.** _____

**29.** A paving crew completes 10,000 feet of a road every 3 days. Approximately how many days will it take to pave a 7-mile stretch of road? (1 mile = 5280 feet)

**29.** _____

**30.** A model airplane has a wingspan of 8 inches. The actual wingspan of the plane it represents is 38 feet. If the model's fuselage is 12 inches long, how long is the fuselage of the actual plane?

**30.** _____

## Chapter 6 PERCENT

### 6.1     Basics of Percent

| Learning Objectives | |
| --- | --- |
| 1 | Learn the meaning of percent. |
| 2 | Write percents as decimals. |
| 3 | Write decimals as percents. |
| 4 | Understand 100%, 200%, and 300%. |
| 5 | Use 50%, 10%, and 1%. |

### Key Terms

Use the vocabulary terms listed below to complete each statement in exercises 1–3.

**percent**            **ratio**        **decimals**

1.    To compare two quantities that have the same type of units, use a _____.

2.    _____ means per one hundred.

3.    _____ represent parts of a whole.

### Objective 1    Learn the meaning of percent.

*Write as a percent.*

1.    43 people out of 100 drive small cars.       1. _____

2.    The tax is $8 per $100.       2. _____

3.    The cost for labor was $45 for every $100 spent to manufacture an item.       3. _____

4.    38 out of 100 planes departed on time.       4. _____

### Objective 2    Write percents as decimals.

*Write each percent as a decimal.*

5.    42%       5. _____

6.    310%       6. _____

7.    4%       7. _____

8.    10%       8. _____

**9.**    2.5%

**10.**   0.025%

**11.**   0.256%

**9.** _____

**10.** _____

**11.** _____

## Objective 3    Write decimals as percents.

_Write each decimal as a percent._

**12.**   0.30

**13.**   0.2

**14.**   0.07

**15.**   0.564

**16.**   4.93

**17.**   5.5

**18.**   0.036

**12.** _____

**13.** _____

**14.** _____

**15.** _____

**16.** _____

**17.** _____

**18.** _____

## Objective 4    Understand 100%, 200%, and 300%.

_Fill in the blanks._

**19.**    100% of $19 is _____.

**20.**   200% of 170 miles is _____.

**21.**   300% of $76 is _____.

**19.** _____

**20.** _____

**21.** _____

**22.** 100% of 12 dogs is _____.          **22.** _____

**23.** 200% of $520 is _____.             **23.** _____

**24.** 300% of $250 is _____.             **24.** _____

**Objective 5    Use 50%, 10%, and 1%.**

*Fill in the blanks.*

**25.** 50% of 250 signs is _____          **25.** _____

**26.** 10% of 100 years is _____.         **26.** _____

**27.** 50% of 48 copies is _____.         **27.** _____

**28.** 10% of 4920 televisions is _____.  **28.** _____

**29.** 1% of 400 homes is _____.          **29.** _____

**30.** 1% of $98 is _____.                **30.** _____

# Chapter 6 PERCENT

## 6.2   Percents and Fractions

| Learning Objectives |
| --- |
| **1**    Write percents as fractions. |
| **2**    Write fractions as percents. |
| **3**    Use the table of percent equivalents. |

## Key Terms

Use the vocabulary terms listed below to complete each statement in exercises 1–2.

> **percent**      **lowest terms**

1. A fraction is in _____ when its numerator and denominator have no common factor other than 1.

2. A _____ can be written as a fraction with 100 in the denominator.

## Objective 1   Write percents as fractions.

*Write each percent as a fraction or mixed number in lowest terms.*

1. 12%

2. 86%

3. 62.5%

4. 43.6%

5. $16\frac{2}{3}\%$

6. $22\frac{2}{9}\%$

7. 0.5%

8. 0.04%

1. _____

2. _____

3. _____

4. _____

5. _____

6. _____

7. _____

8. _____

**9.**    140%                                  **9.** _____

**10.**    275%                                 **10.** _____

**Objective 2     Write fractions as percents.**

*Write each fraction or mixed number as a percent. Round percents to the nearest tenth, if necessary.*

**11.**    $\dfrac{7}{10}$                                  **11.** _____

**12.**    $\dfrac{81}{100}$                              **12.** _____

**13.**    $\dfrac{12}{25}$                              **13.** _____

**14.**    $\dfrac{64}{75}$                              **14.** _____

**15.**    $\dfrac{47}{50}$                              **15.** _____

**16.**    $\dfrac{5}{9}$                                  **16.** _____

**17.**    $3\frac{4}{5}$                                  **17.** _____

**18.**    $2\frac{3}{4}$

**18.** _____

**19.**    $7\frac{2}{5}$

**19.** _____

**20.**    $4\frac{1}{3}$

**20.** _____

**Objective 3**    **Use the table of percent equivalents.**

*Complete this chart. Round decimals to the nearest thousandth and percents to the nearest tenth, if necessary.*

| Fraction | Decimal | Percent | |
|---|---|---|---|
| **21.** $\frac{1}{2}$ | _____ | _____ | **21.** _____ |
| **22.** _____ | 0.125 | _____ | **22.** _____ |
| **23.** $\frac{1}{4}$ | _____ | _____ | **23.** _____ |
| **24.** $\frac{5}{8}$ | _____ | _____ | **24.** _____ |
| **25.** _____ | _____ | 87.5% | **25.** _____ |

**26.**   $\frac{3}{8}$         _____   _____          **26.** _____

**27.**   _____   _____   $33\frac{1}{3}\%$          **27.** _____

**28.**   $\frac{2}{5}$         _____   _____          **28.** _____

**29.**   _____   0.325   _____          **29.** _____

**30.**   $\frac{2}{3}$         _____   _____          **30.** _____

# Chapter 6 PERCENT

## 6.3    Using the Percent Proportion and Identifying the Components in a Percent Problem

| Learning Objectives |
| --- |
| 1    Learn the percent proportion. |
| 2    Solve for an unknown value in a percent proportion. |
| 3    Identify the percent. |
| 4    Identify the whole. |
| 5    Identify the part. |

### Key Terms

Use the vocabulary terms listed below to complete each statement in exercises 1–3.

**percent proportion**          **whole**          **part**

1.    The _____ in a percent problem is the entire quantity.

2.    The _____ in a percent problem is the portion being compared with the whole.

3.    Part is to whole as percent is to 100 is called the _____.

### Objective 1    Learn the percent proportion.

1.    Write the percent proportion                1. _____


### Objective 2    Solve for an unknown value in a percent proportion.

*Use the percent proportion to solve for the unknown value. Round to the nearest tenth, if necessary. If the answer is a percent, be sure to include a percent sign.*

2.    part = 30, percent = 25                2. _____


3.    part = 160, percent = 20               3. _____


4.    part = 18, percent = 150               4. _____


5.    whole = 48, percent = 25               5. _____

**6.**    whole = 25, percent = 14

6. _____

**7.**    whole = 50, percent = 175

7. _____

**8.**    part = 12, whole = 50

8. _____

**9.**    part = 75, whole = 1500

9. _____

**10.**    part = 160, whole = 120

10. _____

## Objective 3    Identify the percent.

*Identify the percent. Do not try to solve for any unknowns.*

**11.**    83% of what number is 21.5?

11. _____

**12.**    36 is 72% of what number?

12. _____

*Identify the percent in each application problem. Do not try to solve for any unknowns.*

**13.**    A chemical is 42% pure. Of 800 grams of the chemical, how much is pure?

13. _____

**14.**    Sales tax of $8 is charged on an item costing $200. What percent of sales tax is charged?

14. _____

**15.**    17% of Tom's check of $340 is withheld. How much is withheld?

15. _____

**16.**   A team won 12 of the 18 games it played. What          **16.** _____
          percent of its games did it win?

## Objective 4   Identify the whole.

*Identify the whole. Do not try to solve for any unknowns.*

**17.**   71 is what percent of 384?                             **17.** _____

**18.**   0.68% of 487 is what number?                           **18.** _____

**19.**   What is 14% of 78?                                     **19.** _____

*Identify the whole in each application problem. Do not try to solve for any unknowns.*

**20.**   In one storm, Springbrook got 15% of the season's      **20.** _____
          snowfall. Springbrook's total snowfall for that
          season was 30 inches. How many inches of snow fell
          in that one storm?

**21.**   In one state, the sales tax is 8%. On a purchase, the   **21.** _____
          amount of tax was $26. Find the cost of the item
          purchase.

**22.**   In an election, 68% of the registered voters actually   **22.** _____
          voted. If there are 12,452 voters, how many people
          voted?

## Objective 5   Identify the part.

*Identify the part. Do not try to solve for any unknowns.*

**23.**   29.81 is what percent of 508?                          **23.** _____

**24.**     16.74 is 11.9% of what number?           **24.** _____

**25.**     What number is 12.4% of 1408?          **25.** _____

*Identify the part, then set up the percent proportion in each application problem. Do not try to solve for any unknowns.*

**26.**     In a one-day storm, Odentown received 0.3% of the     **26.** _____
              season's total rainfall. Odentown received 4 inches
              of rain on that day. How many inches of rain fell
              during the season?

**27.**     A hatchery is notified that 7% of its shipment of     **27.** _____
              baby salmon did not arrive healthy. Of 1500 salmon
              shipped, how many did not arrive healthy?

**28.**     There are 720 quarts of grape juice in a vat holding a     **28.** _____
              total of 2400 quarts of fruit juice. What percent of
              the vat is grape juice?

**29.**     A teacher of English literature found that 15% of the     **29.** _____
              students' papers are handed in late. If there are 40
              students in a class, how many papers will be handed
              in late?

**30.**     Payroll deductions are 35% of Jason's gross pay. If     **30.** _____
              his deductions total $350, what is his gross pay?

## Chapter 6 PERCENT

**6.4    Using Proportions to Solve Percent Problems**

| **Learning Objectives** |
| :--- |
| 1    Use the percent proportion to find the part. |
| 2    Find the whole using the percent proportion. |
| 3    Find the percent using the percent proportion. |

**Key Terms**

Use the vocabulary terms listed below to complete each statement in exercises 1–2.

       **cross products**      **percent proportion**

1.   Solve a proportion using _____.

2.   The equation $\dfrac{\text{part}}{\text{whole}} = \dfrac{\text{percent}}{100}$ is called the _____.

**Objective 1    Use the percent proportion to find the part.**

*Use the percent proportion to find the part. Round to the nearest tenth, if necessary.*

1.   20% of 1400                 1. _____

2.   9% of 42                     2. _____

*Use multiplication to find the part. Round to the nearest tenth, if necessary.*

3.   175% of 50                3. _____

4.   39.4% of 300             4. _____

5.   0.7% of 3500             5. _____

*Solve each application problem. Round to the nearest tenth, if necessary.*

6.   A library has 330 visitors on Saturday, 20% of     6. _____
      whom are children. How many are children?

**7.** Bonnie Rae spent 15% of her savings on textbooks. If her savings were $560, find the amount that she spent on textbooks.

**7.** _____

**8.** A survey at an intersection found that of 2200 drivers, 43% were wearing seat belts. How many drivers in the survey were wearing seat belts?

**8.** _____

**9.** A family of four with a monthly income of $2100 spends 90% of its earnings and saves the balance. How much does the family save in one month?

**9.** _____

**10.** In the last election, 74% of the eligible people actually voted. If there were 7844 voters, how many people were eligible?

**10.** _____

## Objective 2    Find the whole using the percent proportion.

*Use the percent proportion to find the whole. Round to the nearest tenth, if necessary.*

**11.** 15 is 5% of what number?

**11.** _____

**12.** 36% of what number is 75?

**12.** _____

**13.** 550 is 110% of what number?

**13.** _____

**14.** 4.6% of what number is 69?

**14.** _____

**15.**  24.5 is 0.7% of what number?                    **15.** _____

*Solve each application problem. Round to the nearest tenth, if necessary.*

**16.**  Michael Elders owns stock worth $4250, which is       **16.** _____
17% of the value of his investments. What is the
value of his investments?

**17.**  This year, there are 960 scholarship applications,    **17.** _____
which is 120% of the number of applications last
year. Find the number of applications last year.

**18.**  Kathy Wicklund's overtime pay is $420, which is       **18.** _____
12% of her total pay. What is her total pay?

**19.**  In one chemistry class, 60% of the students passed.   **19.** _____
If 90 students passed, how many students were in the
class?

**20.**  On campus this semester there are 2028 married        **20.** _____
students, which is 26% of the total enrollment. Find
the total enrollment.

**Objective 3   Find the percent using the percent proportion.**

*Use the percent proportion to find the whole. Round to the nearest tenth, if necessary.*

**21.**   What percent of 8000 is 4?                           **21.** _____

**22.** 7 is what percent of 280?

**22.** _____

**23.** 650 is what percent of 13?

**23.** _____

**24.** What percent of 4.5 is 3.9?

**24.** _____

**25.** 550 is what percent of 1000?

**25.** _____

*Solve each application problem. Round to the nearest tenth, if necessary.*

**26.** In one shipment, 695 out of 27,800 crates were damaged. What percent of the crates were damaged?

**26.** _____

**27.** G&G Pharmacy has a total payroll of $89,350, of which $19,657 goes towards employee fringe benefits. What percent of the total payroll goes to fringe benefits?

**27.** _____

**28.** Vera's Antique Shoppe says that of its 5100 items in stock, 4233 are just plain junk, while the rest are antiques. What percent of the number of items in stock is antiques?

**28.** _____

**29.** In a motor cross, the leader has completed 108.8 miles of the 128-mile course. What percent of the total course has she completed?

**29.** _____

**30.** This month's class goal for Easy Writer Pen Company is 1,844,500 ballpoint pens. If 239,785 pens have been sold, what percent of the goal has been reached?

**30.** _____

# Chapter 6 PERCENT

### 6.5    Using the Percent Equation

| Learning Objectives |
| --- |
| 1      Use the percent equation to find the part. |
| 2      Find the whole using the percent equation. |
| 3      Find the percent using the percent equation. |

### Key Terms

Use the vocabulary terms listed below to complete each statement in exercises 1–2.

**percent equation**              **percent**

1.    A number written with a _____ sign means "divided by 100".

2.    The _____ is part $=$ percent $\cdot$ whole.

### Objective 1    Use the percent equation to find the part.

*Find the part using the percent equation. Round to the nearest tenth, if necessary.*

1.    70% of 920                          1. _____

2.    9% of 240                           2. _____

3.    140% of 76                          3. _____

4.    12.4% of 8100                       4. _____

5.    0.4% of 350                         5. _____

6.    125% of 76                          6. _____

*Solve each application problem. Round to the nearest tenth or cent, if necessary.*

7.    A gardener has 56 clients, 25% of whom are         7. _____
      residential. Find the number that are residential.

**8.** The total in sales at Hill's Market last month was $87,428. If the profit was $1\frac{1}{2}$ % of the sales, how much was the profit?

8. _____

**9.** The sales tax rate in New York City is 8.375%. How much is the sales tax on a computer that costs $1200?

9. _____

**10.** A pair of shoes is marked 20% off. If the original price was $56, how much is the discount?

10. _____

**Objective 2   Find the whole using the percent equation.**

*Find the whole using the percent equation. Round to the nearest tenth, if necessary.*

**11.** 64 is 40% of what number?

11. _____

**12.** 75% of what number is 1125?

12. _____

**13.** $12\frac{1}{2}$ % of what number is 270?

13. _____

**14.** 75 is $6\frac{1}{4}$ % of what number?

14. _____

**15.** 35 is 153% of what number?

15. _____

**16.**     170% of what number is 1462?           **16.** _____

*Solve each application problem. Round to the nearest tenth or cent, if necessary.*

**17.**     A tank of an industrial chemical is 25% full. The       **17.** _____
        tank now contains 160 gallons. How many gallons
        will it contain when it is full?

**18.**     Greg has completed 37.5% of the units needed for a     **18.** _____
        degree. If he has completed 45 units, how many are
        needed for a degree?

**19.**     One day last week, 5% of the employees in a        **19.** _____
        company were absent. If 25 employees were absent,
        how many employees are there?

**20.**     Over three months, the NASDAQ composite stock      **20.** _____
        index dropped about 13.25%, or 330 points. What
        was its value at the beginning of the three month
        period? (Round your answer to the nearest dollar.)

**Objective 3     Find the percent using the percent equation.**

*Find the percent using the percent equation. Round to the nearest tenth, if necessary.*

**21.**     15 is what percent of 75?              **21.** _____

**22.**     What percent of 250 is 112.5?         **22.** _____

**23.**     What percent of 160 is 8?            **23.** _____

**24.**     What percent of 90 is 1.35?                        **24.** _____


**25.**     What percent of 18 is 44?                          **25.** _____


**26.**     What percent of 27 is 90?                          **26.** _____


*Solve each application problem. Round to the nearest tenth or cent, if necessary.*

**27.**     The Robinson family earns $2800 per month and       **27.** _____
            saves $700 per month. What percent of the income is
            saved?


**28.**     The Hogan family drove 145 miles of their 500-mile  **28.** _____
            vacation. What percent of the total number of miles
            did they drive?


**29.**     Jane eats 1500 calories a day. If she eats 350 calories  **29.** _____
            for breakfast, what percent of her daily calories is
            her breakfast?


**30.**     A house costs $225,000. The Lees paid $45,000 as a  **30.** _____
            down payment. What percent of the cost of the
            house is their down payment?

## Chapter 6 PERCENT

### 6.6    Solving Application Problems with Percent

| **Learning Objectives** |  |
| --- | --- |
| 1 | Find sales tax. |
| 2 | Find commissions. |
| 3 | Find the discount and sale price. |
| 4 | Find the percent of change. |

### Key Terms

Use the vocabulary terms listed below to complete each statement in exercises 1–4.

**sales tax**                **commission**            **discount**

**percent of increase or decrease**

1.    _____ is a percent of the dollar value of total sales paid to a salesperson.

2.    In a _____ problem, the increase or decrease is a percent of the original amount.

3.    The percent of the total sales charged as tax is called the _____.

4.    The percent of the original price that is deducted from the original price is called the _____.

### Objective 1    Find sales tax.

*Find the amount of sales tax and the total cost. Round answers to the nearest cent, if necessary.*

|  | **Amount of sale** | **Tax Rate** |  |
| --- | --- | --- | --- |
| 1. | $50 | 7% | 1. Tax _____<br><br>Total _____ |
| 2. | $350 | 6.5% | 2. Tax _____<br><br>Total _____ |
| 3. | $67 | 9% | 3. Tax _____<br><br>Total _____ |

*Find the sales tax rate. Round answers to the hundredth, if necessary.*

| | Amount of sale | Amount of Tax | | |
|---|---|---|---|---|
| **4.** | $450 | $36 | **4.** _____ |
| **5.** | $215 | $10.75 | **5.** _____ |
| **6.** | $78 | $1.17 | **6.** _____ |

*Solve each application problem. Round money answers to the nearest cent, if necessary.*

**7.** A television set sells for $750 plus 8% sales tax.          **7.** _____
Find the price of the set including sales tax.

**8.** A gold bracelet costs $1300 not including a sales tax       **8.** _____
of $71.50. Find the sales tax rate.

**Objective 2    Find commissions.**

*Find the commission earned. Round answers to the nearest cent, if necessary.*

| | Amount of sale | Rate of Commission | | |
|---|---|---|---|---|
| **9.** | $6225 | 2.5% | **9.** _____ |
| **10.** | $156,000 | 3% | **10.** _____ |

**11.**    $75,000                  4%                      **11.** _____

*Find the rate of commission. Round answers to the hundredth, if necessary.*

|     | **Amount of Sale** | **Amount of Commission** |     |
| --- | --- | --- |
| **12.** | $3200 | $480 | **12.** _____ |
| **13.** | $5783 | $231.32 | **13.** _____ |
| **14.** | $25,000 | $3750 | **14.** _____ |

*Solve each application problem. Round money answers to the nearest cent, if necessary.*

**15.**  Nicole had sales of $18,306 in the month of October.   **15.** _____
If her rate of commission is 12%, find the amount of
commission that she earned.

**16.**  A business property has just been sold for   **16.** _____
$1,692,804. The real estate agent selling the property
earned a commission of $42,320.10. Find the rate of
commission.

## Objective 3    Find the discount and sale price.

*Find the amount of discount and the amount paid after the discount. Round money answers to the nearest cent, if necessary.*

| | Original price | Rate of Discount | | |
|---|---|---|---|---|
| 17. | $200 | 15% | **17. Discount** _____ | |
| | | | **Amount paid** _____ | |
| 18. | $595.80 | 20% | **18. Discount** _____ | |
| | | | **Amount paid** _____ | |
| 19. | $205.50 | 5% | **19. Discount** _____ | |
| | | | **Amount paid** _____ | |
| 20. | $24.95 | 60% | **20. Discount** _____ | |
| | | | **Amount paid** _____ | |

*Solve each application problem. Round money answers to the nearest cent, if necessary.*

**21.**    Mike Lee can purchase a new car at 8% below        **21.** _____
window sticker price. Find the amount he can save
on a car with a window sticker price of $17,608.

**22.**    A "Super 35% Off Sale" begins today. What is the        **22.** _____
price of a hair dryer normally priced at $15?

**23.** Geishe's Shoes sells shoes at 33% off the regular price. Find the price of a pair of shoes normally priced at $54, after the discount is given.

**23.** _____

## Objective 4    Find the percent of change.

*Solve each application problem. Round to the nearest tenth of a percent, if necessary.*

**24.** Enrollment in secondary education courses increased from 1900 students last semester to 2280 students this semester. Find the percent of increase.

**24.** _____

**25.** The number of days employees of Prodex Manufacturing Company were absent from their jobs decreased from 96 days last month to 72 days this month. Find the percent of decrease.

**25.** _____

**26.** The earnings per share of Amy's Cosmetic Company decreased from $1.20 to $0.86 in the last year. Find the percent of decrease.

**26.** _____

**27.** The membership of Pleasant Acres Golf Club was 320 two years ago. The membership is now 740. Find the percent of increase in the two years.

**27.** _____

**28.** The price of a certain model of calculator was $33.50 five years ago. This calculator now costs $18.75. Find the percent of decrease in the price in the last five years.

**28.** _____

**29.** In 1980, there were approximately 3,612,000 births in the U.S. In 2002, there were approximately 4,022,000 births in the U.S. Find the percent of increase.

**29.** _____

**30.** One day in 2008, the Dow Jones Industrial Average dropped from about 12,635 to 12,265. Find the percent of decrease.

**30.** _____

# Chapter 6 PERCENT

### 6.7    Simple Interest

| **Learning Objectives** |
| --- |
| 1    Find the simple interest on a loan. |
| 2    Find the total amount due on a loan. |

### Key Terms

Use the vocabulary terms listed below to complete each statement in exercises 1–5.

> **interest      interest formula      simple interest      principal**
>
> **rate of interest**

1.    The charge for money borrowed or loaned, expressed as a percent, is called

_____.

2.    A fee paid for borrowing or lending money is called _____.

3.    The formula $I = p \cdot r \cdot t$ is called the _____.

4.    Use the formula $I = p \cdot r \cdot t$ to compute the amount of _____ due on a loan.

5.    The amount of money borrowed or loaned is called the _____.

### Objective 1    Find the simple interest on a loan.

*Find the interest. Round to the nearest cent, if necessary.*

| | **Principal** | **Rate** | **Time in Years** | |
| --- | --- | --- | --- | --- |
| 1. | $400 | 2% | 3 | 1. _____ |
| 2. | $80 | 5% | 1 | 2. _____ |
| 3. | $5280 | 8% | 5 | 3. _____ |
| 4. | $780 | 10% | $2\frac{1}{2}$ | 4. _____ |

| | Principal | Rate | Time in Years | | |
|---|---|---|---|---|---|
| 5. | $360 | 6% | $1\frac{1}{2}$ years | 5. | _____ |
| 6. | $620 | 16% | $1\frac{1}{4}$ years | 6. | _____ |

*Find the interest. Round to the nearest cent, if necessary.*

| | Principal | Rate | Time in Months | | |
|---|---|---|---|---|---|
| 7. | $200 | 16% | 3 | 7. | _____ |
| 8. | $500 | 11% | 9 | 8. | _____ |
| 9. | $820 | 3% | 18 | 9. | _____ |
| 10. | $522 | 8% | 21 | 10. | _____ |
| 11. | $2000 | 12% | 39 | 11. | _____ |
| 12. | $14,400 | 7% | 7 | 12. | _____ |

*Solve each application problem. Round to the nearest cent, if necessary.*

13. Diane lends $6500 for 18 months at 12%. How much interest will she earn?

13. _____

14. A mother lends $6500 to her daughter for 15 months and charges 9% interest. Find the interest charged on the loan.

14. _____

**15.** Kareem invests $1500 at 16% for 6 months. What amount of interest will he earn?

**15.** _____

**16.** Darla deposits $680 at 14% for 1 year. How much interest will she earn?

**16.** _____

**17.** A savings account pays $2\frac{1}{2}$% interest per year. How much interest will be earned on $850 invested for 3 years?

**17.** _____

## Objective 2    Find the total amount due on a loan.

*Find the total amount due on each loan. Round to the nearest cent, if necessary.*

|      | Principal | Rate | Time |      |
|------|-----------|------|------|------|
| **18.** | $200 | 11% | 1 year | **18.** _____ |
| **19.** | $3000 | 5% | 6 months | **19.** _____ |
| **20.** | $1500 | 8% | 18 months | **20.** _____ |
| **21.** | $900 | 10% | $2\frac{1}{2}$ years | **21.** _____ |
| **22.** | $6000 | 7% | 5 months | **22.** _____ |

| | **Principal** | **Rate** | **Time** | | |
|---|---|---|---|---|---|
| **23.** | $15,400 | 16% | 5 years | **23.** | _____ |
| **24.** | $18,200 | 7% | 8 months | **24.** | _____ |
| **25.** | $30,900 | 16% | 4 months | **25.** | _____ |

*Solve each application problem. Round to the nearest cent, if necessary.*

**26.** A loan of $1500 will be paid back with 12% interest at the end of 27 months. Find the total amount due.

**26.** _____

**27.** An employee credit union pays 7% interest. If Mario deposits $2100 in his account for $\frac{1}{3}$ year and makes no withdrawals or further deposits, find the total amount in Mario's account after that time.

**27.** _____

**28.** An investor deposits $7000 at 16% for 2 years. If there are no withdrawals or further deposits, find the total amount in the account after 2 years.

**28.** _____

**29.** Mary Ann borrows $1200 at 10% for 3 months. Find the total amount due.

**29.** _____

**30.** Cheri borrowed $45,000 for 9 months at 11.2% interest. Find the total amount she must repay.

**30.** _____

## Chapter 6 PERCENT

### 6.8    Compound Interest

| **Learning Objectives** |
| --- |
| 1    Understand compound interest. |
| 2    Understand compound amount. |
| 3    Find the compound amount. |
| 4    Use a compound interest table. |
| 5    Find the compound amount and the amount of compound interest. |

### Key Terms

Use the vocabulary terms listed below to complete each statement in exercises 1–3.

    **compound interest**        **compound amount**        **compounding**

1.    Interest paid on principal plus past interest is called _____.

2.    The total amount in an account, including compound interest and the original principal, is called the _____.

3.    When the amount of interest is computed based on the principal plus the past interest, use a process called _____.

**Objective 1    Understand compound interest.**
**Objective 2    Understand compound amount.**

1.    Belinda deposited $2000 in an account earning 5%        1. _____
    annually. How much is in the account at the end of
    the first year?

2.    If Belinda makes no withdrawals, how much money        2. _____
    is in her account at the end of two years?

3.    If Belinda makes no withdrawals, how much money        3. _____
    is in her account at the end of three years? How
    much interest has she earned in total?            _____

## Objective 3    Find the compound amount.

*Find the compound amount given the following deposits. Interest is compounded annually. Round to the nearest cent, if necessary.*

**4.**    $7000 at 5% for 3 years                    **4.** _____

**5.**    $4500 at 4% for 2 years                    **5.** _____

**6.**    $3200 at 7% for 2 years                    **6.** _____

**7.**    $24,600 at 5% for 4 years                  **7.** _____

**8.**    $8000 at 8% for 4 years                    **8.** _____

**9.**    $3200 at 6% for 2 years                    **9.** _____

**10.**   $1200 at 2% for 3 years                    **10.** _____

## Objective 4    Use a compound interest table.

*Use the table for compound interest to find the compound amount. Interest is compounded annually. Round to the nearest cent, if necessary.*

| Time Periods | 3.00% | 3.50% | 4.00% | 4.50% | 5.00% | 5.50% | 6.00% | 8.00% | Time Periods |
|---|---|---|---|---|---|---|---|---|---|
| 1 | 1.0300 | 1.0350 | 1.0400 | 1.0450 | 1.0500 | 1.0550 | 1.0600 | 1.0800 | 1 |
| 2 | 1.0609 | 1.0712 | 1.0816 | 1.0920 | 1.1025 | 1.1130 | 1.1236 | 1.1664 | 2 |
| 3 | 1.0927 | 1.1087 | 1.1249 | 1.1412 | 1.1576 | 1.1742 | 1.1910 | 1.2597 | 3 |
| 4 | 1.1255 | 1.1475 | 1.1699 | 1.1925 | 1.2155 | 1.2388 | 1.2625 | 1.3605 | 4 |
| 5 | 1.1593 | 1.1877 | 1.2167 | 1.2462 | 1.2763 | 1.3070 | 1.3382 | 1.4693 | 5 |
| 6 | 1.1941 | 1.2293 | 1.2653 | 1.3023 | 1.3401 | 1.3788 | 1.4185 | 1.5869 | 6 |
| 7 | 1.2299 | 1.2723 | 1.3159 | 1.3609 | 1.4071 | 1.4547 | 1.5036 | 1.7138 | 7 |
| 8 | 1.2668 | 1.3168 | 1.3686 | 1.4221 | 1.4775 | 1.5347 | 1.5938 | 1.8509 | 8 |
| 9 | 1.3048 | 1.3629 | 1.4233 | 1.4861 | 1.5513 | 1.6191 | 1.6895 | 1.9990 | 9 |
| 10 | 1.3439 | 1.4106 | 1.4802 | 1.5530 | 1.6289 | 1.7081 | 1.7908 | 2.1589 | 10 |
| 11 | 1.3842 | 1.4600 | 1.5395 | 1.6229 | 1.7103 | 1.8021 | 1.8983 | 2.3316 | 11 |
| 12 | 1.4258 | 1.5111 | 1.6010 | 1.6959 | 1.7959 | 1.9012 | 2.0122 | 2.5182 | 12 |

**11.** $1000 at 6% for 4 years              11. _____

**12.** $4000 at 5% for 9 years              12. _____

**13.** $7500 at 6% for 7 years              13. _____

**14.** $60 at 5.5% for 2 years              14. _____

**15.** $48 at 8% for 3 years                15. _____

**16.**  $8428.17 at $4\frac{1}{2}$% for 6 years                    **16.** _____

**17.**  $10,422.75 at $5\frac{1}{2}$% for 12 years               **17.** _____

**18.**  $24,600 at 5% for 4 years                                **18.** _____

## Objective 5   Find the compound amount and the amount of compound interest.

*Find the compound amount and the compound interest. Round to the nearest cent, if necessary. Use the table for compound interest in your text book to find the compound amount. Interest is compounded annually.*

| | Principal | Rate | Time in Years | |
|---|---|---|---|---|
| **19.** | $1000 | $3\frac{1}{2}$% | 7 | **19.** Amount _____  Interest_____ |
| **20.** | $8500 | 6% | 12 | **20.** Amount _____  Interest_____ |
| **21.** | $12,800 | $5\frac{1}{2}$% | 9 | **21.** Amount _____  Interest_____ |
| **22.** | $9150 | 8% | 8 | **22.** Amount _____  Interest_____ |
| **23.** | $21,400 | $4\frac{1}{2}$% | 11 | **23.** Amount _____  Interest_____ |

| | **Principal** | **Rate** | **Time in Years** | |
|---|---|---|---|---|
| **24.** | $45,000 | 4% | 4 | **24.** Amount _____ |
| | | | | Interest _____ |
| **25.** | $78,000 | 3% | 12 | **25.** Amount _____ |
| | | | | Interest _____ |
| **26.** | $8000 | 3% | 12 | **26.** Amount _____ |
| | | | | Interest _____ |
| **27.** | $1000 | 4.5% | 6 | **27.** Amount _____ |
| | | | | Interest _____ |
| **28.** | $8500 | 8% | 12 | **28.** Amount _____ |
| | | | | Interest _____ |

*Solve each application problem. Use the table for compound interest in your text book to find the compound amount. Interest is compounded annually. Round to the nearest cent, if necessary.*

**29.** Scott lends $9000 to the owner of a new restaurant. He will be repaid at the end of 6 years at 8% interest compounded annually. Find how much he will be repaid and how much interest he will earn.

**29.**
Amount _____

Interest _____

**30.** Michelle invests $2500 in a health spa. She will be repaid at the end of 5 years at 6% interest compounded annually. Find how much she will be repaid and how much interest she will earn.

**30.**
Amount _____

Interest _____

# Chapter 7 GEOMETRY

### 7.1    Lines and Angles

| Learning Objectives |
|---|
| 1    Identify and name lines, line segments, and rays. |
| 2    Identify parallel and intersecting lines. |
| 3    Identify and name angles. |
| 4    Classify angles as right, acute, straight, or obtuse. |
| 5    Identify perpendicular lines. |
| 6    Identify complementary angles and supplementary angles and find the measure of complement or supplement of a given angle. |
| 7    Identify congruent angles and vertical angles and use this knowledge to find the measures of angles. |
| 8    Identify corresponding angles and alternate interior angles and use this knowledge to find the measures of angles. |

### Key Terms

Use the vocabulary terms listed below to complete each statement in exercises 1–19.

| | | | |
|---|---|---|---|
| point | line | line segment | ray |
| parallel lines | intersecting lines | angle | |
| degrees | straight angle | right angle | |
| acute angle | obtuse angle | perpendicular lines | |
| complementary angles | supplementary angles | | |
| congruent angles | vertical angles | | |
| corresponding angles | alternate interior angles | | |

1.    A _____ is a part of a line that has one endpoint and which extends infinitely in one direction.

2.    Two lines that intersect to form a right angle are _____.

3.    An angle whose measure is between 90° and 180° is an _____.

4.    A _____ is a location in space.

5.    Two rays with a common endpoint form an _____.

6.    A set of points that form a straight path that extends infinitely in both directions is called a _____.

7.    An angle that measures less than 90° is called an _____.

8.    Angles are measured using _____.

**9.**   Two lines in the same plane that never intersect are _____.

**10.**  A part of a line with two endpoints is a _____.

**11.**  Two lines that cross at one point are _____.

**12.**  An angle whose measure is exactly 90° is a _____.

**13.**  The nonadjacent angles formed by two intersecting lines are called
_____.

**14.**  Angles whose measures are equal are called _____.

**15.**  Two angles whose measures sum to 180° are _____.

**16.**  Two angles whose measures sum to 90° are _____.

**17.**  When two parallel lines are cut by a transversal, the angles between the parallel
lines on opposite sides of the transversal are called
_____.

**18.**  When two parallel lines are cut by a transversal, the angles in the same relative
position with regard to the parallel lines and the transversal are called
_____.

**19.**  An angle whose measure is exactly 180° is a _____.

**Objective 1    Identify and name lines, line segments, and rays.**

*Identify each figure as a line, line segment, or ray, and name it.*

**1.**   C
            \
             \
              D

**1.** _____

**2.**
         ↑
       A ●
         |
         |
       B ●

**2.** _____

**3.**  ←●———●→
        E    F

**3.** _____

Name:                       Date:

Instructor:               Section:

**Objective 2    Identify parallel and intersecting lines.**

*Label each pair of lines as appearing to be **parallel** or **intersecting**.*

**4.**

**4.** _____

**5.**

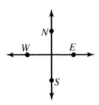

**5.** _____

**6.**

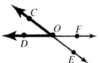

**6.** _____

**Objective 3    Identify and name angles.**

*Name each angle drawn with darker rays by using the three-letter form of identification.*

**7.**

**7.** _____

**8.**

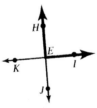

**8.** _____

**9.**

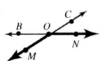

**9.** _____

**10.**

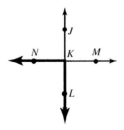

**10.** _____

## Objective 4   Classify angles as right, acute, straight, or obtuse.

*Label each angle as* **acute**, **right**, **obtuse**, *or* **straight**.

**11.**

**11.** _____

**12.**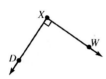

**12.** _____

**13.**

**13.** _____

**14.**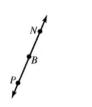

**14.** _____

## Objective 5   Identify perpendicular lines.

*Label each pair of lines as appearing to be* **parallel**, **perpendicular**, *or* **neither**.

**15.**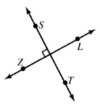

**15.** _____

**16.**

**16.** _____

**17.**

**17.** _____

**18.**

18. _____

**Objective 6**   **Identify complementary angles and supplementary angles and find the measure of complement or supplement of a given angle.**

*Identify each pair of complementary angles.*

**19.**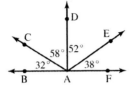

19. _____

**20.**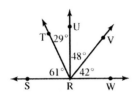

20. _____

*Identify each pair of supplementary angles.*

**21.**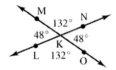

21. _____

**22.**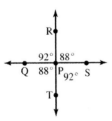

22. _____

*Find the complement and supplement of each angle.*

**23.**   16°

**23.** complement _____

supplement _____

**24.**   72°

**24.** complement _____

supplement _____

**Objective 7**   **Identify congruent angles and vertical angles and use this knowledge to find the measures of angles.**

*In each of the following, identify the angles that are congruent. Then find the measure of each angle.*

**25.**   ∠POQ measures 34°.

**25.** _____

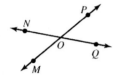

**26.**   ∠VLC measures 157°.

**26.** _____

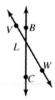

**Objective 8**   **Identify corresponding angles and alternate interior angles and use this knowledge to find the measures of angles.**

*In each figure, line m is parallel to line n. List the corresponding angles and the alternate interior angles. Then find the measure of each angle.*

**27.**   ∠4 measures 100°.

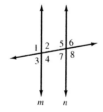

**27.**

Corresponding angles:

_____

Alternate interior angles:

_____

Angle measures:

_____

**28.**    ∠6 measures 125°.

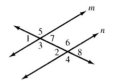

**28.**
Corresponding angles:

_____

Alternate interior angles:

_____

Angle measures:


_____


**29.**    7 measures 37°.

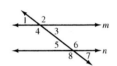

**29.**
Corresponding angles:

_____

Alternate interior angles:


_____

Angle measures:


_____


**30.**    ∠3 measures 44°.

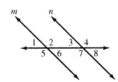

**30.**
Corresponding angles:

_____

Alternate interior angles:


_____

Angle measures:


_____

# Chapter 7 GEOMETRY

## 7.2    Rectangles and Squares

| **Learning Objectives** |
| --- |
| 1     Find the perimeter and area of a rectangle. |
| 2     Find the perimeter and area of a square. |
| 3     Find the perimeter and area of a composite figure. |

## Key Terms

Use the vocabulary terms listed below to complete each statement in exercises 1–4.

**perimeter     area       rectangle     square**

1. The number of square units in a region is called the _____ of the region.

2. A four-sided figure with four right angles is called a _____.

3. The distance around the outside edges of a figure is called the _____ of the figure.

4. A rectangle with four equal sides is called a _____.

## Objective 1    Find the perimeter and area of a rectangle.

*Find the perimeter and area of each rectangle.*

1. 4 centimeters by 8 centimeters         1. _____

2. 17 inches by 12 inches         2. _____

3. 1 centimeter by 17 centimeters         3. _____

4. 14.5 meters by 3.2 meters         4. _____

5. $4\frac{1}{2}$ yards by $6\frac{1}{2}$ yards         5. _____

6. 87.2 feet by 33 feet         6. _____

7. 37.4 centimeters by 103.2 centimeters         7. _____

*Solve each application problem.*

**8.** A picture frame measures 20 inches by 30 inches. Find the perimeter and area of the frame.

**8.** _____

**9.** A lot is 114 feet by 212 feet. County rules require that nothing be built on land within 12 feet of any edge of the lot. Find the area on which you cannot build.

**9.** _____

**10.** A room is 14 yards by 18 yards. Find the cost to carpet this room if carpet costs $23 per square yard.

**10.** _____

**Objective 2    Find the perimeter and area of a square.**

*Find the perimeter and area of each square with the given side.*

**11.** 9 meters

**11.** _____

**12.** 9.2 yards

**12.** _____

**13.** 7.8 feet

**13.** _____

**14.** 13 feet

**14.** _____

**15.** $1\frac{2}{5}$ inches

**15.** _____

**16.** 8.2 km

**16.** _____

**17.** 3.1 cm

**17.** _____

**18.** 7.4 inches

**18.** _____

**19.**     $4\frac{2}{3}$ miles

**20.**     21 m

## Objective 3    Find the perimeter and area of a composite figure.

*Find the perimeter and area of each figure. All angles that appear to be right angles are, in fact, right angles.*

**21.**

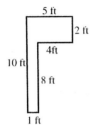

21. _____

**22.**

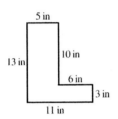

22. _____

**23.**

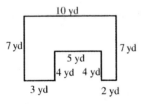

23. _____

**24.**

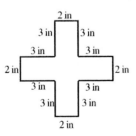

24. _____

**25.**

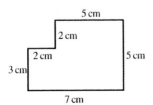

25. _____

**26.**

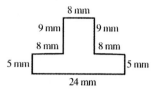

8 mm
9 mm   9 mm
8 mm   8 mm
5 mm   5 mm
24 mm

**26.** _____

**27.**

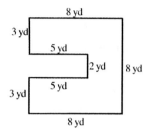

8 yd
3 yd
5 yd
2 yd
8 yd
3 yd
5 yd
8 yd

**27.** _____

**28.**

15 m
12 m
5 m
29 m   5 m
5 m
12 m
15 m

**28.** _____

**29.**

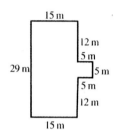

12 cm
9 cm
16 cm
3 cm
7 cm
9 cm

**29.** _____

**30.**

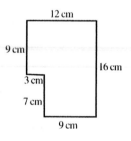

10 ft
18 ft   20 ft
40 ft   10 ft
14 ft
32 ft
44 ft

**30.** _____

# Chapter 7 GEOMETRY

### 7.3     Parallelograms and Trapezoids

| **Learning Objectives** |
| :--- |
| 1       Find the perimeter and area of a parallelogram. |
| 2       Find the perimeter and area of a trapezoid. |

### Key Terms

Use the vocabulary terms listed below to complete each statement in exercises 1–4.

      **perimeter     area        parallelogram      trapezoid**

1.    A _____ is a four-sided figure with both pairs of opposite sides parallel and equal in length.

2.    A _____ is a four-sided figure with exactly one pair of parallel sides.

3.    The formula $P = 2 \cdot l + 2 \cdot w$ is the formula for the _____ of a rectangle.

4.    Square the length of a side of a square to find the _____ of a square.

### Objective 1    Find the perimeter and area of a parallelogram.

*Find the perimeter of each parallelogram.*

1.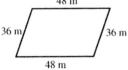

                                                       1. _____

2.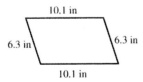

                                                       2. _____

3.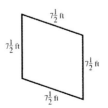

                                                       3. _____

**4.**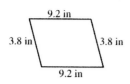

**4.** _____

**5.**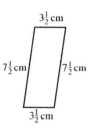

**5.** _____

*Find the area of each parallelogram.*

**6.**

**6.** _____

**7.**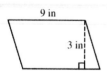

**7.** _____

**8.**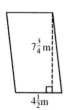

**8.** _____

**9.**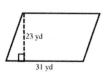

**9.** _____

**10.**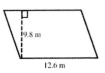

**10.** _____

**11.**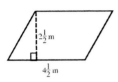

**11.** _____

*Solve each application problem.*

**12.** A parallelogram has a height of 3.2 meters and a base of 4.6 meters. Find the area.

**12.** _____

**13.** A parallelogram has a height of $15\frac{1}{2}$ feet and a base of 20 feet. Find the area.

**13.** _____

**14.** A swimming pool is in the shape of a parallelogram with a height of 9.6 meters and base of 12 meters. Find the cost of a solar pool cover that sells for $5.10 per square meter.

**14.** _____

**15.** An auditorium stage has a hardwood floor that is shaped like a parallelogram, having a height of 30 feet and a base of 40 feet. If a company charges $0.65 per square foot to refinish floors, find the cost of refinishing the stage floor.

**15.** _____

## Objective 2    Find the perimeter and area of a trapezoid.

*Find the perimeter of each figure.*

**16.**

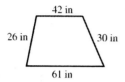

**16.** _____

**17.**

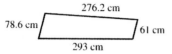

**17.** _____

**18.**

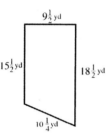

**18.** _____

*Find the area of each figure.*

**19.**

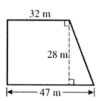

**19.** _____

**20.**

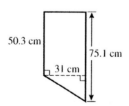

**20.** _____

**21.**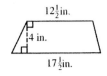

**21.** _____

**22.**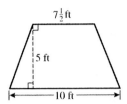

**22.** _____

**23.**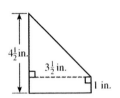

**23.** _____

**24.**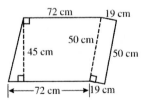

**24.** _____

**25.**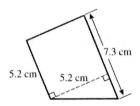

**25.** _____

**26.**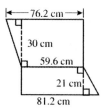

**26.** _____

**27.**

**27.** _____

*Solve each application problem.*

**28.** The lobby in a resort hotel is in the shape of a
trapezoid. The height of the trapezoid is 52 feet and
the bases are 47 feet and 59 feet. Carpet that costs
$2.75 per square foot is to be laid in the lobby. Find
the cost of the carpet.

**28.** _____

**29.** The backyard of a new home is shaped like a
trapezoid, having a height of 35 feet and bases of 90
feet and 110 feet. Find the cost of planting a lawn in
the yard if the landscaper charges $0.20 per square
foot.

**29.** _____

**30.** A hot tub is in the shape shown. Find the cost of a
cover for the hot tub at a cost of $9.70 per square
foot. Angles that appear to be right angles are indeed
right angles.

**30.** _____

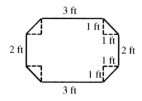

# Chapter 7 GEOMETRY

### 7.4    Triangles

| Learning Objectives |
|---|
| 1    Find the perimeter of a triangle. |
| 2    Find the area of a triangle. |
| 3    Given the measures of two angles in a triangle, find the measure of the third angle. |

### Key Terms

Use the vocabulary terms listed below, along with the figure, to complete each statement in exercises 1–3.

**base**          **height**          **triangle**

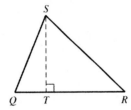

1.    A figure with exactly three sides is called a _____.

2.    In the figure, $\overline{QR}$ is the _____ of $\Delta QRS$.

3.    In the figure, $\overline{ST}$ is the _____ of $\Delta QRS$.

### Objective 1    Find the perimeter of a triangle.

*Find the perimeter of each triangle.*

1.

    1.    _____

2.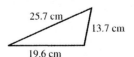

    2.    _____

**3.**

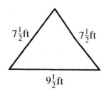

**3.** _____

**4.**

**4.** _____

**5.**

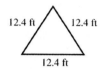

**5.** _____

**6.**

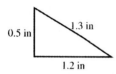

**6.** _____

**7.** A triangle with sides $2\frac{1}{2}$ feet, 3 feet, and $5\frac{1}{4}$ feet

**7.** _____

**8.** A triangle with two equal sides of 3.6 centimeters and the third side 4.1 centimeters

**8.** _____

**9.** A triangle with three equal sides each 5.9 meters

**9.** _____

**10.**   A triangle with sides $13\frac{1}{8}$ inches, $11\frac{3}{4}$ inches, and      **10.** _____

   $14\frac{1}{2}$ inches.


## Objective 2    Find the perimeter of a triangle.

*Find the area of each triangle.*

**11.**

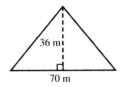

**11.** _____

**12.**

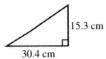

**12.** _____

**13.**

**13.** _____

**14.**

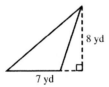

**14.** _____

**15.**

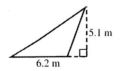

**15.** _____

**16.**

**16.** _____

*Find the shaded area in each figure.*

**17.**

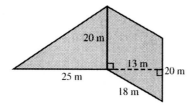

**17.** _____

**18.**

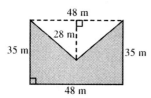

**18.** _____

**19.**

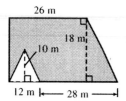

**19.** _____

**20.**

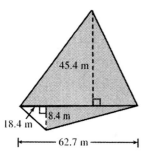

**20.** _____

**21.**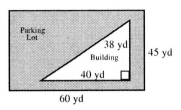

**21.** _____

**22.** If a painter charged $4.06 per square meter to paint the front of the house shaded below, how much would he charge? All angles that appear to be right angles are right angles.

**22.** _____

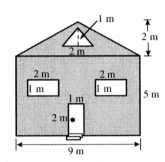

**Objective 3**    **Given the measures of two angles in a triangle, find the measure of the third angle.**

*The measures of two angles of a triangle are given. Find the measure of the third angle.*

**23.**    100°, 63°

**23.** _____

**24.**    60°, 60°

**24.** _____

**25.**    37°, 62°

**25.** _____

**26.**    49°, 72°

**26.** _____

**27.**    51°, 78°

**27.** _____

**28.**    $87°, 13°$                            **28.** _____

**29.**    $90°, 45°$                            **29.** _____

**30.**    $76°, 76°$                            **30.** _____

# Chapter 7 GEOMETRY

### 7.5    Circles

| **Learning Objectives** |
| --- |
| 1    Find the radius and diameter of a circle. |
| 2    Find the circumference of a circle. |
| 3    Find the area of a circle. |
| 4    Become familiar with Latin and Greek prefixes used in math terminology. |

### Key Terms

Use the vocabulary terms listed below to complete each statement in exercises 1–5.

> **circle**       **radius**       **diameter**       **circumference**       **π (pi)**

1.  The _____ is the distance from the center of a circle to any point on the circle.

2.  The _____ of a circle is the distance around the circle.

3.  A figure whose points lie the same distance from a fixed center point is called a
    _____.

4.  The ratio of the circumference to the diameter of any circle equals _____.

5.  The _____ of a circle is a segment connecting two points on a circle and passing through the center.

### Objective 1    Find the radius and diameter of a circle.

*Find the diameter or radius in each circle.*

1.
$d = ?$
$r = 43$ m

1. _____

2.
$d = ?$
$r = 64$ ft

2. _____

**3.**

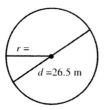

**3.** _____

**4.**

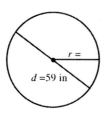

**4.** _____

**5.** The diameter of a circle is 8 feet. Find its radius.

**5.** _____

**6.** The radius of a circle is 2.7 centimeters. Find its diameter.

**6.** _____

**7.** The diameter of a circle is $12\frac{1}{2}$ yards. Find its radius

**7.** _____

**Objective 2    Find the circumference of a circle.**

_Find the circumference of each circle. Use 3.14 as an approximation for_ $\pi$. _Round each answer to the nearest tenth._

**8.**

**8.** _____

**9.**

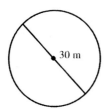

**9.** _____

**10.**     A circle with a diameter of $4\frac{3}{4}$ inches         **10.** _____

**11.**     A circle with a radius of 4.5 yards         **11.** _____

**12.**     A circle with a radius of $\frac{3}{4}$ mile         **12.** _____

*Solve each application problem.*

**13.**     How far does a point on the tread of a tire move in     **13.** _____
one turn if the diameter of the tire is 60 centimeters?

**14.**     If you swing a ball held at the end of a string 3     **14.** _____
meters long, how far will the ball travel on each
turn?

**Objective 3**     **Find the area of a circle.**

*Find the area of each circle. Use 3.14 as an approximation for $\pi$. Round each answer to the nearest tenth.*

**15.**     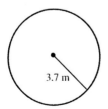         **15.** _____

3.7 m

**16.**

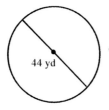

44 yd

**16.** _____

**17.**    A circle with diameter of $5\frac{1}{3}$ yards

**17.** _____

**18.**    A circle with diameter of 9.8 centimeters

**18.** _____

*Find the area of the shaded region. Use 3.14 as an approximation for $\pi$. Round each answer to the nearest tenth.*

**19.**

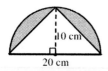

10 cm

20 cm

**19.** _____

**20.**

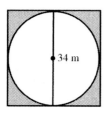

34 m

**20.** _____

**21.**

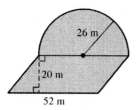

26 m

20 m

52 m

**21.** _____

**22.**

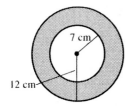

**22.** _____

**23.**

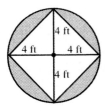

**23.** _____

*Solve each application problem.*

**24.** Find the area of a circular pond that has a diameter of 12.6 meters.

**24.** _____

**25.** Find the cost of sod, at $1.80 per square foot, for the following playing field. Round the answer to the nearest cent.

**25.** _____

**Objective 4   Become familiar with Latin and Greek prefixes used in math terminology.**

*Write one math term and one nonmathematical term that use each prefix listed below. (Answers will vary.)*

**26.**   *tri-* (three)

**26.** _____

**27.**   *poly*-(many)

**27.** _____

**28.**   *milli*-(thousand)

**28.** _____

**29.**   *oct*-(eight)

**29.** _____

**30.**   *uni*-(one)

**30.** _____

## Chapter 7 GEOMETRY

### 7.6    Volume and Surface Area

| **Learning Objectives** |
| --- |
| 1    Find the volume of a rectangular solid. |
| 2    Find the volume of a sphere. |
| 3    Find the volume of a cylinder. |
| 4    Find the volume of a cone and a pyramid. |
| 5    Find the surface area of a rectangular solid. |
| 6    Find the surface area of a cylinder. |

### Key Terms

Use the vocabulary terms listed below to complete each statement in exercises 1–7.

> **volume**        **rectangular solid**      **sphere**        **cylinder**        **cone**
>
> **pyramid**        **surface area**

1.   A _____ is a box-like solid figure.

2.   A solid figure with two congruent, parallel, circular bases is a
     _____.

3.   A _____ is a ball-like solid figure.

4.   _____ is a measure of the space inside a solid shape.

5.   A solid figure whose base is a square or a rectangle and whose faces (sides) are
     triangles is called a _____.

6.   A solid figure with only one base, and that base is a circle, is called a
     _____.

7.   The area on the surface of a three-dimensional object is its _____.

### Objective 1    Find the volume of a rectangular solid.

*Find the volume of each rectangular solid. Round answers to the nearest tenth, if necessary.*

1.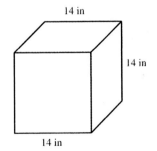

1.   _____

**2.**

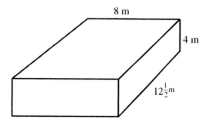

8 m

4 m

$12\frac{1}{2}$ m

**2.** _____

**3.**

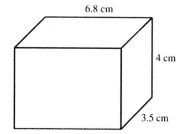

6.8 cm

4 cm

3.5 cm

**4.** _____

*Find the volume of each solid.*

**4.**

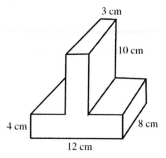

3 cm

10 cm

4 cm

8 cm

12 cm

**4.** _____

**5.**

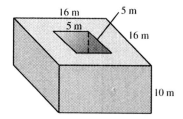

16 m

5 m

5 m

16 m

10 m

(Notice the square hole that goes through the center of the shape.)

**5.** _____

### Objective 2    Find the volume of a sphere.

*Find the volume of each sphere or hemisphere. Use 3.14 as an approximation for π.*
*Round answers to the nearest tenth, if necessary.*

**6.**

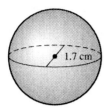

6. _____

**7.**

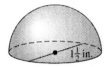

7. _____

**8.**    A sphere with a diameter of $3\frac{1}{4}$ inches.

8. _____

**9.**    A hemisphere with a radius of 11.6 feet.

9. _____

**10.**    A sphere with a radius of 6.8 cm

10. _____

### Objective 3    Find the volume of a cylinder.

*Find the volume of each figure. Use 3.14 as an approximation for π. Round answers to*
*the nearest tenth, if necessary.*

**11.**

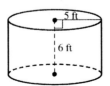

11. _____

**12.**

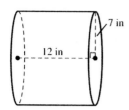

**12.** _____

**13.** A cardboard mailing tube, diameter 5 centimeters and height 25 centimeters

**13.** _____

**14.**

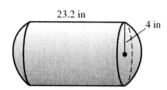

**14.** _____

**15.**

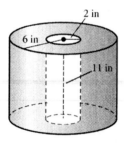

**15.** _____

## Objective 4    Find the volume of a cone and a pyramid.

_Find the volume of each figure. Use 3.14 as an approximation for $\pi$. Round answers to the nearest tenth, if necessary._

**16.**

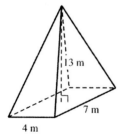

**23.** _____

**17.**

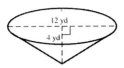

**17.** _____

**18.**

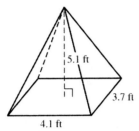

**18.** _____

**19.** Find the volume of a pyramid with square base 42 meters on a side and height 38 meters.

**19.** _____

**20.** Find the volume of a cone with base diameter 3.2 centimeters and height 5.8 centimeters.

**20.** _____

**Objective 5    Find the surface area of a rectangular solid.**

*Find the surface area of each rectangular solid. Round your answers to the nearest tenth.*

**21.**

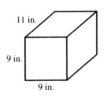

**21.** _____

**22.**

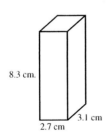

**22.** _____

**23.**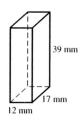

39 mm

17 mm

12 mm

**23.** _____

**24.**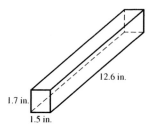

12.6 in.

1.7 in.

1.5 in.

**24.** _____

**25.**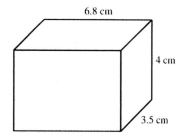

6.8 cm

4 cm

3.5 cm

**25.** _____

**Objective 6   Find the surface area of a cylinder.**

*Find the surface area of each cylinder. Use 3.14 as the approximate value for π. Round your answers to the nearest tenth.*

**26.**

24 in.

8 in.

**26.** _____

**27.**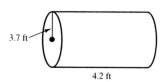

3.7 ft

4.2 ft

**27.** _____

**28.**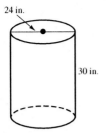

24 in.

30 in.

**28.** _____

**29.**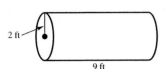

2 ft

9 ft

**29.** _____

**30.**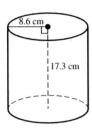

8.6 cm

17.3 cm

**30.** _____

# Chapter 7 GEOMETRY

### 7.7    Pythagorean Theorem

| Learning Objectives |
| --- |
| 1    Find square roots using the square root key on a calculator. |
| 2    Find the unknown length in a right triangle. |
| 3    Solve application problems involving right triangles. |

### Key Terms

Use the vocabulary terms listed below to complete each statement in exercises 1–3.

**hypotenuse**            **legs**            **right triangle**

1.    A triangle with a 90° angle is called a _____.

2.    The side opposite the right angle in a right triangle is called the
      _____ of the triangle.

3.    The two sides of the right angle in a right triangle are called the
      _____ of the triangle.

### Objective 1    Find square roots using the square root key on a calculator.

*Find each square root. Use a calculator with a square root key. Round the answer to the nearest thousandth, if necessary.*

1.    $\sqrt{17}$                                                        1. _____

2.    $\sqrt{27}$                                                        2. _____

3.    $\sqrt{2}$                                                         3. _____

4.    $\sqrt{55}$                                                        4. _____

5.    $\sqrt{75}$                                                        5. _____

6.    $\sqrt{102}$                                                       6. _____

**7.**    $\sqrt{145}$

7. _____

### Objective 2    Find the unknown length in a right triangle.

*Find the unknown length in each right triangle. Use a calculator with a square root key.*
*Round the answer to the nearest tenth, if necessary.*

**8.**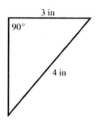

8. _____

**9.**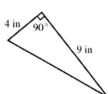

9. _____

**10.**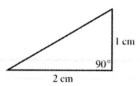

10. _____

**11.**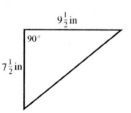

11. _____

**12.**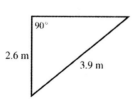

12. _____

**13.**

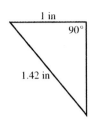

**13.** _____

**14.**

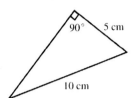

**14.** _____

**15.**

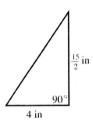

**15.** _____

**16.**

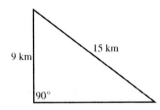

**16.** _____

**17.**

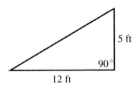

**17.** _____

## Objective 3    Solve application problems involving right triangles.

*Solve each application problem. Draw a diagram if one is not provided. Use a calculator with a square root key. Round the answer to the nearest tenth, if necessary.*

**18.**    Find the length of the loading ramp.

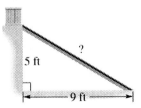

**18.** _____

**19.** Find the unknown length in this roof plan.              **19.** _____

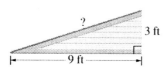

**20.** A boat travels due south from a dock 10 miles and     **20.** _____
then turns and travels due east for 15 miles. How far
is the boat from the dock?

**21.** A boat is pulled into a dock with a rope attached to    **21.** _____
the bottom of the boat. When the boat is 12 feet from
the dock, the length of the rope is 13 feet. How high
is the dock?

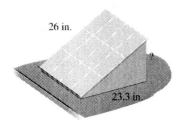

**22.** A solar panel on a roof is 26 inches wide. It is        **22.** _____
mounted on a frame whose base is 23.3 inches long.
How tall is the frame?

**23.** A kite is flying on 50 feet of string. If the horizontal distance of the kite from the person flying it is 40 feet, how far off the ground is the kite?

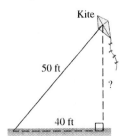

**24.** Find the distance between the centers of the holes in the metal plate.

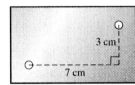

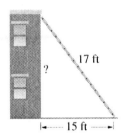

**25.** The base of a 17-ft ladder is located 15 ft from a building. How high up on the building will the ladder reach?

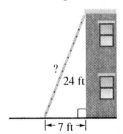

**26.** The base of a ladder is located 7 feet from a building. The ladder reaches 24 feet up the building. How long is the ladder?

**27.**  A truck is stopped 6 feet from a door into a storeroom. If the back of the truck is 4 feet above ground level, how long a ramp is needed to unload the truck?

**27.** _____

**28.**  What is the radius of the circle?

**28.** _____

**29.**  A repairman needs to fix the siding located 18 feet up from the ground on a house. Since there are bushes next to the house. The base of the ladder must be 6.5 feet from the house. How long must the ladder be to reach the repair site?

**29.** _____

**30.**  An access ramp is being built to a door in a building as shown below. Find the length of the ramp.

**30.** _____

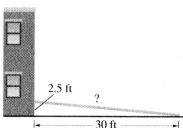

# Chapter 7 GEOMETRY

### 7.8    Congruent and Similar Triangles

| **Learning Objectives** |
| --- |
| 1    Identify corresponding parts of congruent triangles. |
| 2    Prove that triangles are congruent using SAS, SSS, and ASA. |
| 3    Identify corresponding parts in similar triangles. |
| 4    Find the unknown lengths of sides in similar triangles. |
| 5    Solve application problems involving similar triangles. |

### Key Terms

Use the vocabulary terms listed below to complete each statement in exercises 1–4.

    **congruent figures**        **similar figures**

    **congruent triangles**      **similar triangles**

1.    Triangles with the same shape and size are _____.

2.    _____ are identical both in shape and in size.

3.    Triangles with the same shape but not necessarily the same size are

    _____.

4.    _____ have the same shape but are different sizes.

### Objective 1    Identify corresponding parts of congruent triangles.

*Each pair of triangles is congruent. List the corresponding angles and the corresponding sides.*

1.    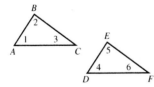                              1.    _____

2.    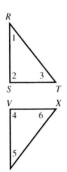                              2.    _____

## Objective 2   Prove that triangles are congruent using SAS, SSS, and ASA.

*Determine which of these methods can be used to prove that each pair of triangles is congruent: Angle-Side-Angle (ASA), Side-Side-Side (SSS), or Side-Angle-Side (SAS).*

**3.**
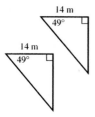

**3.** _____

**4.**
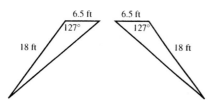

**4.** _____

**5.**
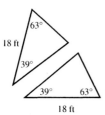

**5.** _____

**6.**
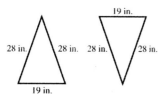

**6.** _____

**7.**
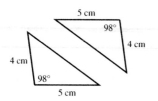

**7.** _____

**8.**
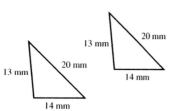

**8.** _____

**9.**

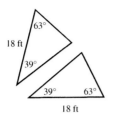

**9.** _____

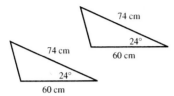

**10.**

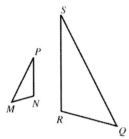

**10.** _____

**Objective 3    Identify the corresponding parts in similar triangles.**

*Name the corresponding angles and the corresponding sides in each pair of similar triangles.*

**11.**

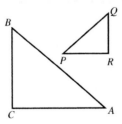

**11.** _____

**12.**

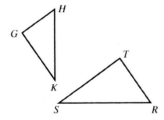

**12.** _____

**13.**

**13.** _____

243

**14.**

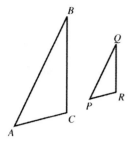

**14.** _____

*Write the ratio for each pair of corresponding sides in the similar triangles shown below. Write the ratios as fractions in lowest terms.*

**15.** $\dfrac{XY}{NM}, \dfrac{XZ}{NL}, \dfrac{YZ}{ML}$

**15.** _____

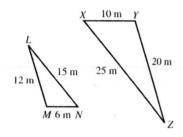

**16.** $\dfrac{QR}{BA}, \dfrac{SR}{CA}, \dfrac{SQ}{CB}$

**16.** _____

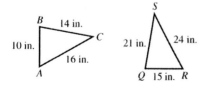

17.   $\dfrac{LM}{PS}$ ; $\dfrac{MK}{ST}$ ; $\dfrac{LK}{PT}$                    17. _____

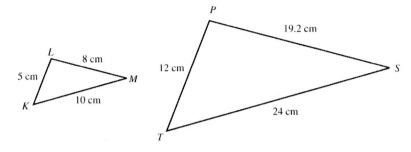

**Objective 4    Find the unknown lengths of sides in similar triangles.**

*Find the unknown lengths in each pair of similar triangles. Note that the figures may not be drawn to scale.*

18.   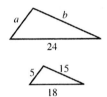                    18. _____

19.   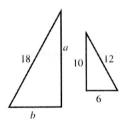                    19. _____

20.   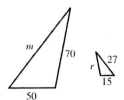                    20. _____

**21.**

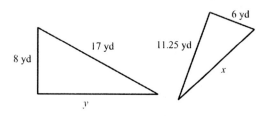

**21.** _____

Find the unknown length in the following. Round the answer to the nearest tenth, if necessary. Note: When a line is drawn parallel to one side of a triangle, the smaller triangle that is formed will be similar to the original triangle.

**22.**

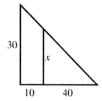

**22.** _____

Find the perimeter of each triangle. Assume the triangles are similar.

**23.**

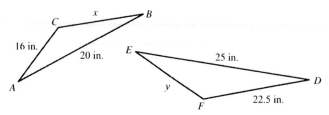

**23.** ABC_____

    DEF _____

**24.**

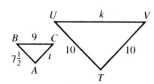

**24.** ABC_____

    TUV_____

Name: _____     Date: _____

Instructor: _____     Section: _____

## Objective 5     Solve application problems involving similar triangles.

*Solve each application problem.*

25.  A flagpole casts a shadow 52 m long at the same time that a pole 9 m tall casts á shadow 12 m long. Find the height of the flagpole.

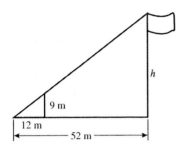

26.  The height of the house shown here can be found by using similar triangles and proportion. Find the height of the house by writing a proportion and solving it.

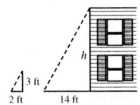

27.  A sailor on the USS Ramapo saw one of the highest waves ever recorded. He used the height of the ship's mast, the length of the deck and similar triangles to find the height of the wave. Using the information in the figure, write a proportion and then find the height of the wave.

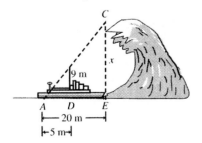

**28.**   The ratio of the rise of a roof to the run of a roof is 5     **28.** _____
to 12. Use this information to find the height of the
roof indicated by *h* in the diagram.

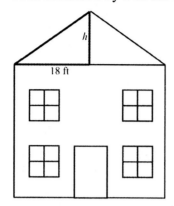

**29.**   Use similar triangles to find the distance *h* across the    **29.** _____
river in the figure.

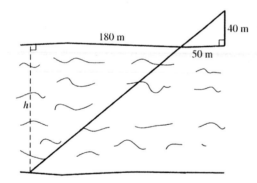

**30.**   Find the height of the tree. (Hint: Since eye-level is     **30.** _____
1.8 m above the ground, first find *y* and then add 1.8
meters for the distance from the ground to eye level.)

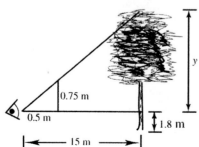

# Chapter 8 STATISTICS

### 8.1 Circle Graphs

| **Learning Objectives** |
| --- |
| 1       Read and understand a circle graph. |
| 2       Use a circle graph. |
| 3       Draw a circle graph. |

### Key Terms

Use the vocabulary terms listed below to complete each statement in exercises 1–2.

      **circle graph**            **protractor**

1.      A _____ shows how a total amount is divided into parts or sectors.

2.      A _____ is a device used to measure the number of degrees in angles or parts of a circle.

### Objective 1     Read and understand a circle graph.

*The circle graph shows the cost of remodeling a kitchen. Use the graph to answer exercises 1–6.*

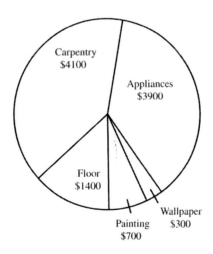

1.      Find the total cost of remodeling the kitchen.            1. _____

2.      What is the largest single expense in remodeling the kitchen?            2. _____

3.      How much less does the wallpaper cost than painting?            3. _____

**4.** What fraction of the total cost of remodeling are the appliances?

4. _____

**5.** Find the ratio of the cost of wallpaper to the cost of the floor.

5. _____

**6.** Find the ratio of the cost of painting to the cost of the floor.

6. _____

## Objective 2    Use a circle graph.

*The circle graph shows the number of students enrolled in certain majors at a college. Use the graph to answer exercises 7–12.*

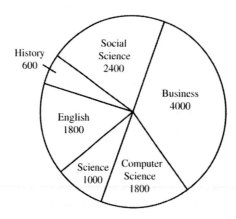

**7.** Which major has the most number of students enrolled?

7. _____

**8.** Find the ratio of the number of business majors to the total number of students.

8. _____

**9.** Find the ratio of the number of English majors to the total number of students.

9. _____

**10.** Find the ratio of the number of science majors to the number of English majors.

10. _____

**11.** Find the ratio of the number of history majors to the number of social science majors.

11. _____

**12.** Find the ratio of the number of computer science majors to the number of business majors.

**12.** _____

*The circle graph shows the expenses involved in keeping a sales force on the road. Each expense item is expressed as a percent of the total sales force cost of $950,000. Find the number of dollars of expense for each category in Problems 13–18. Then, use those answers and the circle graph to answer Problems 19–21.*

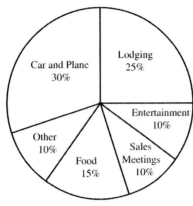

**13.** Car and plane

**13.** _____

**14.** Lodging

**14.** _____

**15.** Entertainment

**15.** _____

**16.** Sales meetings

**16.** _____

**17.** Food

**17.** _____

**18.** Other

**18.** _____

**19.** What is the ratio of food expense to sales meetings expense?

**19.** _____

**20.** What percent of the total expenses is spent on food and entertainment?

**20.** _____

**21.** What is the ratio of car and plane expenses to lodging expenses?

**21.** _____

Name:                                    Date:
Instructor:                              Section:

*The circle graph shows the enrollment by major at a small college. The total enrollment at the college is 3200 students. Use the circle graph to answer questions 22–27.*

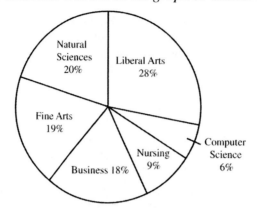

22.   What is the most popular major at the college?          22. _____

23.   What major has the fewest students?                     23. _____

24.   How many business and computer science majors           24. _____
      are there in all?

25.   Find the ratio of business majors to computer           25. _____
      science majors.

26.   Find the ratio of natural science majors to liberal arts   26. _____
      majors.

27.   What percent of the students are either nursing or      27. _____
      natural sciences majors?

## Objective 3    Draw a circle graph.

*Use the given information to draw a circle graph.*

28.    Jensen Manufacturing Company has its annual sales divided into five categories as follows.

| Item | Annual Sales |
|------|-------------|
| Parts | $20,000 |
| Hand tools | 80,000 |
| Bench tools | 100,000 |
| Brass fittings | 140,000 |
| Cabinet hardware | 60,000 |

(a)    Find the total sales for a year.

**a.** _____

(b)    Find the percent of the total sales for each item.

**b.** parts _____

hand tools _____

bench tools _____

brass fittings _____

hardware _____

(c)    Find the number of degrees in a circle graph for each item.

**c.** parts _____

hand tools _____

bench tools _____

brass fittings _____

hardware _____

(d)     Make a circle graph showing this information.

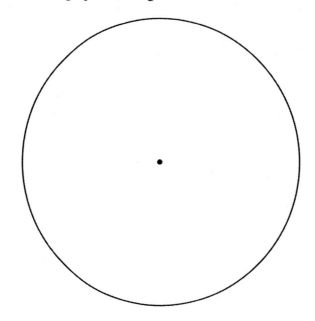

29.     A book publisher had 30% of its sales in mysteries, 15% in biographies, 10% in cookbooks, 25% in romance novels, 15% in science, and the rest in business books.

    (a)     Find the number of degrees in a circle graph for each type of book.

    **a.** mysteries_____

           biographies _____

           cookbooks_____

           romance _____

           science _____

           business_____

(b)    Draw a circle graph showing this
       information.

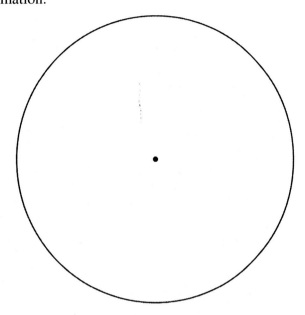

**30.**    A family recorded its expenses for a year, with the following results.

| Item | Percent of Total |
|------|------------------|
| Housing | 40% |
| Food | 20% |
| Automobile | 14% |
| Clothing | 8% |
| Medical | 6% |
| Savings | 8% |
| Other | 4% |

(a)    Find the number of degrees in a circle graph
       for each item.

**a.** housing _____

food _____

automobile _____

clothing _____

medical _____

savings _____

other _____

(b)     Draw a circle graph showing this
        information.

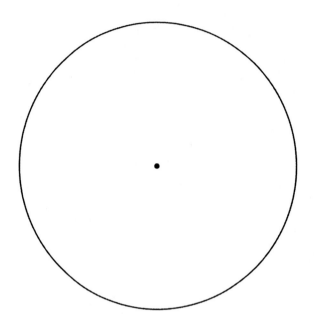

# Chapter 8 STATISTICS

### 8.2    Bar Graphs and Line Graphs

| Learning Objectives |
| --- |
| 1     Read and understand a bar graph. |
| 2     Read and understand a double-bar graph. |
| 3     Read and understand a line graph. |
| 4     Read and understand a comparison line graph. |

### Key Terms

Use the vocabulary terms listed below to complete each statement in exercises 1–4.

**bar graph      double-bar graph      line graph      comparison line graph**

1.    A _____ uses dots connected by line to show trends.

2.    A _____ compares two sets of data by showing two sets of bars.

3.    A _____ uses bars of various heights or lengths to show quantity or frequency.

4.    A _____ shows how two sets of data relate to each other by showing a line graph for each item.

### Objective 1    Read and understand a bar graph.

*The bar graph shows the enrollment for the fall semester at a small college for the past five years. Use this graph for problems 1–7.*

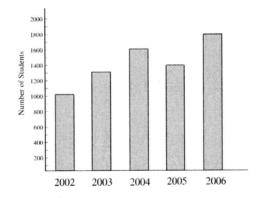

1.    What was the fall enrollment for 2002?              1. _____

2.    What was the fall enrollment for 2004?              2. _____

3.    What was the fall enrollment for 2006?              3. _____

**4.** How many more students were enrolled in 2004 than in 2003?

**4.** _____

**5.** What year had the greatest enrollment?

**5.** _____

**6.** Which year showed a decrease in enrollment?

**6.** _____

**7.** By how many students did the enrollment increase from 2005 to 2006?

**7.** _____

## Objective 2    Read and understand a double-bar graph.

*The double-bar graph shows the enrollment by gender in each class at a small college. Use the double-bar graph for Problems 8–14.*

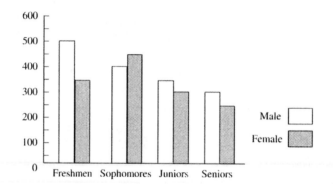

**8.** Which class has a greater female enrollment than male enrollment?

**8.** _____

**9.** How many female freshmen are enrolled?

**9.** _____

**10.** Find the total number of juniors enrolled.

**10.** _____

**11.** Find the ratio of freshmen males to freshmen females.

**11.** _____

**12.** Find the total number of students enrolled.

**12.** _____

**13.** Find the ratio of freshmen students to senior students.

**13.** _____

**14.** Which class has the greatest difference between male students and female students?

**14.** _____

## Objective 3  Read and understand a line graph.

*The line graph gives the value of one share of stock of Microchip Computer Corporation on the first trading day of the month for six consecutive months. Use the line graph for Problems 15–21.*

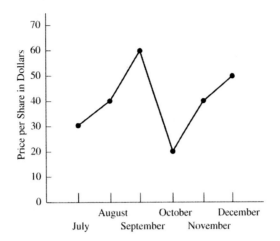

15.  In which month was the value of the stock highest?          15. _____

16.  Find the value of one share on the first trading day          16. _____
     October.

17.  Find the increase in the value of one share from          17. _____
     October to November.

18.  What is the largest monthly decrease in the value of          18. _____
     one share?

19.  Find the ratio of the value of one share on the first          19. _____
     trading day in September to the value of one share
     on the first trading day of October.

20.  Comparing the value of one share on the first trading          20. _____
     day in July to the first trading day in November, has
     the value increased, decreased, or remained
     unchanged?

21.  By how much did the value of one share increase          21. _____
     from July to September?

## Objective 4    Read and understand a comparison line graph.

*The comparison line graph shows annual sales for two different stores for each of the past few years. Use the graph to solve Problems 22–30.*

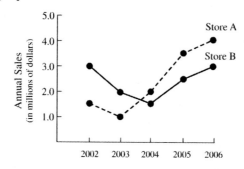

*Find the annual sales for store A in each of the following years.*

**22.**   2005                                    **22.** _____

**23.**   2003                                    **23.** _____

**24.**   2002                                    **24.** _____

*Find the annual sales for store B in each of the following years.*

**25.**   2005                                    **25.** _____

**26.**   2003                                    **26.** _____

**27.**   2002                                    **27.** _____

**28.**   In which years did the sales of store *A* exceed the        **28.** _____
         sales of store *B*?

**29.**   Which year showed the least difference between the          **29.** _____
         sales of store *A* and the sales of store *B*?

**30.**   Find the ratio of the sales of store *A* to the sales of    **30.** _____
         store *B* in 2003.

# Chapter 8 STATISTICS

## 8.3    Frequency Distributions and Histograms

| Learning Objectives | |
|---|---|
| 1 | Understand a frequency distribution. |
| 2 | Arrange data in class intervals. |
| 3 | Read and understand a histogram. |

### Key Terms

Use the vocabulary terms listed below to complete each statement in exercises 1–2.

**frequency distribution        histogram**

1.   A bar graph in which the width of each bar represents a range of number and the height represents the quantity or frequency of items that fall within the interval is called a _____.

2.   A table that includes a column showing each possible number in the data collected is called a _____.

### Objective 1    Understand a frequency distribution.

*The following scores were earned by students on an algebra exam. Use the data to find the tally and the frequency for the given score in problems 1–6.*

| 84 | 90 | 83 | 72 | 84 | 93 | 83 | 90 | 83 |
|----|----|----|----|----|----|----|----|----|
| 90 | 72 | 64 | 90 | 83 | 72 | 83 | 83 | 64 |

1.   64

**1.** tally _____

frequency _____

2.   72

**2.** tally _____

frequency _____

3.   83

**3.** tally _____

frequency _____

4.   84

**4.** tally _____

frequency _____

**5.**    90

     **5.** tally _____

          frequency _____

**6.**    93

     **6.** tally _____

          frequency _____

## Objective 2    Arrange data in class intervals.

*The following list of numbers represents systolic blood pressure of 21 patients. Use the data to find the tally and the frequency for the given score in problems 7–12. Then answer problems 13 and 14.*

| 120 | 98 | 180 | 128 | 143 | 98 | 105 |
|-----|-----|-----|-----|-----|-----|-----|
| 136 | 115 | 190 | 118 | 105 | 180 | 112 |
| 160 | 110 | 138 | 122 | 98 | 175 | 118 |

**7.**    90–109

     **7.** tally _____

          frequency _____

**8.**    110–129

     **8.** tally _____

          frequency _____

**9.**    130–149

     **9.** tally _____

          frequency _____

**10.**    150–169

     **10.** tally _____

          frequency _____

**11.**    179–189

     **11.** tally _____

          frequency _____

**12.**    190–209

     **12.** tally _____

          frequency _____

**13.**    What was the most common range of systolic blood pressure?

     **13.** _____

**14.** What was the least common range of systolic blood pressure?

**14.** _____

*The following list of numbers represents IQ scores of 18 students. Use these numbers to find the tally and the frequency for the given score in problems 15–19. Then answer problems 20 and 21.*

| 98 | 121 | 112 | 99 | 105 | 112 |
|----|-----|-----|----|-----|-----|
| 110 | 100 | 92 | 109 | 104 | 106 |
| 105 | 88 | 92 | 103 | 98 | 118 |

**15.** 80–89

**15.** tally _____

frequency _____

**16.** 90–99

**16.** tally _____

frequency _____

**17.** 100–109

**17.** tally _____

frequency _____

**18.** 110–119

**18.** tally _____

frequency _____

**19.** 120–129

**19.** tally _____

frequency _____

**20.** What was the most common range of IQ scores?

**20.** _____

**21.** What was the least common range of IQ scores

**21.** _____

*For problems 22 and 23, construct a histogram using the given data.*

**22.** The following list of numbers represents systolic
blood pressures of 21 patients. Use intervals 90–109,
110–129, 130–149, 150–169, 170–189, and
190–209.

**22.** _____

| 120 | 98  | 180 | 128 | 143 | 98  | 105 |
|-----|-----|-----|-----|-----|-----|-----|
| 136 | 115 | 190 | 118 | 105 | 180 | 102 |
| 160 | 110 | 138 | 122 | 98  | 175 | 118 |

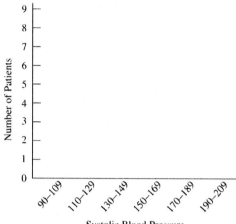

**23.** The following list of numbers represents IQ scores
of 18 students. Use intervals 80–89, 90–99,
100–109, 110–119, and 120–129.

**23.** _____

| 98  | 121 | 112 | 99  | 105 | 112 |
|-----|-----|-----|-----|-----|-----|
| 110 | 100 | 92  | 109 | 104 | 106 |
| 105 | 88  | 92  | 103 | 98  | 118 |

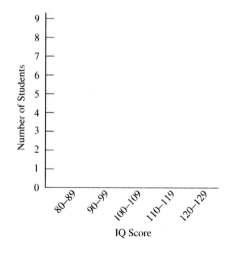

Name:                                    Date:
Instructor:                              Section:

## Objective 3    Read and understand a histogram.

*A local chess club recorded the ages of their members and constructed a histogram. Use the histogram to solve Problems 24–30.*

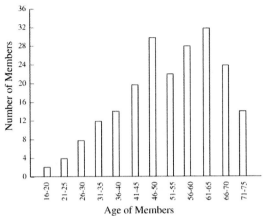

**24.**  The greatest number of members is in which age group?

24. _____

**25.**  The fewest number of members are in which age group?

25. _____

**26.**  Find the number of members 30 years of age or younger.

26. _____

**27.**  Find the number of members 51 years and older.

27. _____

**28.**  How many members are 51–65 years of age?

28. _____

**29.**  How many members are 46–50 years of age?

29. _____

**30.**  Which age range contains the least number of members?

30. _____

Name:                                    Date:
Instructor:                              Section:

# Chapter 8 STATISTICS

## 8.4    Mean, Median, and Mode

| Learning Objectives |
| --- |
| **1**    Find the mean of a list of numbers. |
| **2**    Find a weighted mean. |
| **3**    Find the median. |
| **4**    Find the mode. |

### Key Terms

Use the vocabulary terms listed below to complete each statement in exercises 1–7.

> **mean**        **weighted mean**        **median**        **mode**
>
> **bimodal**        **dispersion**        **range**

1.    The _____ is the variation or spread of the numbers around the mean.

2.    The _____ is the value that occurs most often in a group of values.

3.    A mean calculated so that each value is multiplied by its frequency is called a _____.

4.    The sum of all the values in a data set divided by the number of values in the data set is called the _____.

5.    The middle number in a group of values that are listed from smallest to largest is called the _____.

6.    When two values in a data set occur the same number of times, the data set is called _____.

7.    The difference between the largest value and the smallest value in a set of numbers is called the _____.

### Objective 1    Find the mean of a list of numbers.

*Find the mean for each list of numbers. Round to the nearest tenth, if necessary.*

1.    39, 50, 59, 61, 69, 73, 51, 80                    1. _____

2.    31, 37, 44, 51, 52, 74, 69, 83                    2. _____

3.    62.7, 59.6, 71.2, 65.8, 63.1                      3. _____

**4.**    19900, 23850, 25930, 27710, 29340, 41000          **4.** _____

**5.**    40.1, 32.8, 82.5, 51.2, 88.3, 31.7, 43.7, 51.2     **5.** _____

**6.**    216, 245, 268, 268, 280, 291, 304, 313             **6.** _____

**7.**    2.8, 3.9, 4.7, 5.6, 6.5, 9.1                       **7.** _____

## Objective 2    Find a weighted mean.

*Find the weighted mean for each list of numbers. Round to the nearest tenth, if necessary.*

**8.**

| Value | Frequency |
|-------|-----------|
| 17    | 4         |
| 12    | 5         |
| 15    | 3         |
| 19    | 1         |

**8.** _____

**9.**

| Value | Frequency |
|-------|-----------|
| 13    | 4         |
| 12    | 2         |
| 19    | 5         |
| 15    | 3         |
| 21    | 1         |
| 27    | 5         |

**9.** _____

**10.**

| Value | Frequency |
|-------|-----------|
| 35    | 1         |
| 36    | 2         |
| 39    | 5         |
| 40    | 4         |
| 42    | 3         |
| 43    | 5         |

**10.** _____

**11.**

| Value | Frequency |
|-------|-----------|
| 1     | 2         |
| 2     | 3         |
| 4     | 5         |
| 5     | 7         |
| 6     | 4         |
| 7     | 2         |
| 8     | 1         |
| 9     | 1         |

**11.** _____

*Find the grade point average for each of the following students.  Assume A = 4, B = 3, C = 2, D = 1, F = 0. Round to the nearest tenth, if necessary.*

**12.**

| Units | Grade |
| --- | --- |
| 4 | C |
| 2 | B |
| 5 | C |
| 1 | D |
| 3 | F |

12. _____

**13.**

| Units | Grade |
| --- | --- |
| 3 | C |
| 3 | A |
| 4 | B |
| 5 | B |
| 2 | A |

13. _____

**14.**

| Units | Grade |
| --- | --- |
| 3 | A |
| 4 | B |
| 2 | C |
| 5 | C |
| 2 | D |

14. _____

**15.**

| Units | Grade |
| --- | --- |
| 5 | B |
| 4 | C |
| 3 | B |
| 2 | C |
| 2 | C |

15. _____

**Objective 3    Find the median.**

*Find the median for each list of numbers.*

**16.**    199, 472, 312, 298, 254

16. _____

**17.**    200, 215, 226, 238, 250, 283

17. _____

**18.**    0.002, 0.004, 0.012, 0.008

18. _____

**19.**  389, 464, 521, 610, 654, 672, 682, 712                    **19.** _____

**20.**  43, 69, 108, 32, 51, 49, 83, 57, 64                       **20.** _____

**21.**  21, 32, 27, 23, 25, 29, 22                                **21.** _____

**22.**  1.8, 1.2, 1.1, 1.9, 2.6                                   **22.** _____

**23.**  200, 195, 302, 284, 256, 237, 239, 240                    **23.** _____

**Objective 4    Find the mode.**

*Find the mode for each list of numbers.*

**24.**  32, 43, 57, 43, 59, 43, 57                                **24.** _____

**25.**  4, 9, 3, 4, 7, 3, 2, 3, 9                                 **25.** _____

**26.**  238, 272, 274, 272, 268, 271                              **26.** _____

**27.**  37, 24, 35, 35, 24, 38, 39, 28, 27, 39                    **27.** _____

**28.**  172.6, 199.7, 182.4, 167.1, 172.6, 183.4, 187.6           **28.** _____

**29.**  2, 4, 6, 6, 8, 10, 8, 12, 14, 8                           **29.** _____

**30.**  0.2, 0.7, 0.9, 0.7, 0.5, 0.3, 0.4, 0.7, 0.2               **30.** _____

# Chapter 9 THE REAL NUMBER SYSTEM

### 9.1     Exponents, Order of Operations, and Inequality

| **Learning Objectives** |
| --- |
| 1      Use exponents. |
| 2      Use the rules for order of operations. |
| 3      Use more than one grouping symbol. |
| 4      Know the meanings of $\neq$, $>$, $<$, $\leq$, and $\geq$. |
| 5      Translate word statements to symbols. |
| 6      Write statements that change the direction of inequality symbols. |

### Key Terms

Use the vocabulary terms listed below to complete each statement in exercises 1–3.

     **exponent**          **base**          **exponential expression**

1. A number written with an exponent is an _____ .

2. The _____ is the number that is a repeated factor when written with an exponent.

3. An _____ is a number that indicates how many times a factor is repeated.

### Objective 1    Use exponents.

*Find the value of each exponential expression.*

1.    $3^3$                                 1. _____

2.    $\left(\dfrac{1}{2}\right)^6$                         2. _____

3.    $\left(\dfrac{2}{3}\right)^4$                         3. _____

4.    $(.4)^2$                              4. _____

5.    $(2.4)^3$                           5. _____

## Objective 2  Use the rules for order of operations.

*Find the value of each expression.*

6.     $20 \div 5 - 3 \cdot 1$                 6. _____

7.     $3 \cdot 5^2 - 3 \cdot 7 - 9$              7. _____

8.     $6^2 \div 3^2 - 4 \cdot 3 - 2 \cdot 5$        8. _____

9.     $\dfrac{3 \cdot 15 + 10^2}{12^2 - 8^2}$              9. _____

10.    $\dfrac{10(5-3) - 9(6-2)}{2(4-1) - 2^2}$      10. _____

## Objective 3  Use more than one grouping symbol.

*Find the value of each expression.*

11.    $8\left[14 - 3(9-4)\right]$          11. _____

12.    $19 - 3\left[8(5-2) + 6\right]$      12. _____

13.    $4\left[5 + 2(8-6)\right] + 12$      13. _____

**14.** $3^3\left[(6+5)-2^2\right]$                    **14.** _____

**15.** $2\left[4+2\left(5^2-3\right)\right]$                    **15.** _____

**Objective 4    Know the meanings of $\neq$, $>$, $<$, $\leq$, and $\geq$.**

*Tell whether each statement is* true *or* false.

**16.** $3\cdot 4 \div 2^2 \neq 3$                    **16.** _____

**17.** $3.25 > 3.52$                    **17.** _____

**18.** $2\left[7(4)-3(5)\right]\leq 45$                    **18.** _____

**19.** $4\frac{1}{2}+2\frac{3}{4}<7$                    **19.** _____

**20.** $4 \geq \dfrac{2(3+1)-3(2+1)}{3\cdot 2-1}$                    **20.** _____

**Objective 5    Translate word statements to symbols.**

*Write each word statement in symbols.*

**21.**    Seven equals thirteen minus six.                    **21.** _____

**22.** Seven is greater than the quotient of fifteen and five.

**22.** _____

**23.** The difference between thirty and seven is greater than twenty.

**23.** _____

**24.** Five times the sum of two and nine is less than one hundred six.

**24.** _____

**25.** Twenty is greater than or equal to the product of two and seven.

**25.** _____

**Objective 6    Write statements that change the direction of inequality symbols.**

*Write each statement with the inequality symbol reversed.*

**26.** $\dfrac{2}{7} \le \dfrac{4}{5}$

**26.** _____

**27.** $\dfrac{3}{4} > \dfrac{2}{3}$

**27.** _____

**28.** $12 \ge 8$

**28.** _____

**29.** $.002 > .0002$

**29.** _____

**30.** $\dfrac{3}{8} \le \dfrac{3}{7}$

**30.** _____

# Chapter 9 THE REAL NUMBER SYSTEM

## 9.2    Variables, Expressions, and Equations

| Learning Objectives |
| --- |
| 1    Evaluate algebraic expressions, given values for the variables. |
| 2    Translate phrases from words to algebraic expressions. |
| 3    Identify solutions of equations. |
| 4    Translate sentences to equations. |
| 5    Distinguish between *expressions* and *equations*. |

### Key Terms

Use the vocabulary terms listed below to complete each statement in exercises 1–4.

variable        algebraic expression        equation        solution

1.    An _____ is a statement that says two expressions are equal.

2.    A _____ is a symbol, usually a letter, used to represent an unknown number.

3.    A collection of numbers, variables, operation symbols, and grouping symbols is an_____.

4.    Any value of a variable that makes an equation true is a _____ of the equation.

### Objective 1    Evaluate algebraic expressions, given values for the variables.

*Find the value of each expression if $x = 2$ and $y = 4$.*

1.    $9x - 3y + 2$                                    1. _____

2.    $.3(8x + 2y)$                                    2. _____

3.    $\dfrac{3x}{4} - \dfrac{3y}{2}$                                    3. _____

**4.** $\dfrac{2x+3y}{3x-y+2}$

**4.** _____

**5.** $\dfrac{3y^2+2x^2}{5x+y^2}$

**5.** _____

**6.** $\dfrac{3x+y^2+2}{2x+3y}$

**6.** _____

**Objective 2    Translate phrases from words to algebraic expressions.**

*Write each word phrase as an algebraic expression. Use x as the variable.*

**7.**   One more than three times a number

**7.** _____

**8.**   17 less than nine times a number

**8.** _____

**9.**   Ten times a number, added to 21

**9.** _____

**10.**   The difference between twice a number and 7

**10.** _____

**11.**   11 fewer than eight times a number

**11.** _____

**12.**   Half a number subtracted from two-thirds of the number

**12.** _____

**Objective 3    Identify solutions of equations.**

*Decide whether the given number is a solution of the equation.*

**13.**   $6b+2(b+3)=14;\ 2$

**13.** _____

**14.**   $5 + 3x^2 = 19;\ 2$                    **14.** _____

**15.**   $\dfrac{m+2}{3m-10} = 1;\ 8$                    **15.** _____

**16.**   $\dfrac{x^2-7}{x} = 6;\ 2$                    **16.** _____

**17.**   $3y + 5(y-5) = 7;\ 4$                    **17.** _____

**18.**   $\dfrac{4-x}{x+2} = \dfrac{7}{5};\ \dfrac{1}{2}$                    **18.** _____

**Objective 4    Translate sentences to equations.**

*Write each word sentence as an equation. Use x as the variable.*

**19.**   The sum of five times a number and two is 23.                    **19.** _____

**20.**   Ten divided by a number is two more than the number.                    **20.** _____

**21.**   The product of six and five more than a number is nineteen.                    **21.** _____

**22.**   The quotient of twenty-four and a number is the difference between the number and two.                    **22.** _____

**23.**     Seven times a number subtracted from 61 is 13 plus the number.

**23.** _____

**24.**     Four times a number is equal to two more than three times the number.

**24.** _____

**Objective 5**    **Distinguish between _expressions_ and _equations_.**

_Identify each as an_ **expression** _or an_ **equation**.

**25.**     $4x + 2y + 7$

**25.** _____

**26.**     $y^2 - 4y - 3$

**26.** _____

**27.**     $\dfrac{x+3}{15} = 2x$

**27.** _____

**28.**     $y^2 - 7y + 4 = 0$

**28.** _____

**29.**     $\dfrac{x+4}{5}$

**29.** _____

**30.**     $8x = 2y$

**30.** _____

# Chapter 9 THE REAL NUMBER SYSTEM

### 9.3    Real Numbers and the Number Line

| Learning Objectives |
| --- |
| 1    Classify numbers and graph them on number lines. |
| 2    Tell which of two real numbers is less than the other. |
| 3    Find the opposite of a real number. |
| 4    Find the absolute value of a real number. |

## Key Terms

Use the vocabulary terms listed below to complete each statement in exercises 1–13.

| natural numbers | whole numbers | number line | opposite |
| --- | --- | --- | --- |
| integers | negative number | positive number | |
| rational number | set-builder notation | | coordinate |
| irrational number | real numbers | absolute value | |

1.   The set {0, 1, 2, 3, …} is called the set of _____.

2.   The _____of a number is the same distance from 0 on the number line as the original number, but located on the other side of 0.

3.   The whole numbers together with their opposites and 0 are called _____.

4.   The set { 1, 2, 3, …} is called the set of _____.

5.   The _____of a number is its distance from 0 on the number line.

6.   A _____ shows the ordering of the real numbers on an infinite line.

7.   A real number that is not a rational number is called a(n) _____.

8.   The number that corresponds to a point on the number line is the _____ of that point.

9.   A number located to the left of 0 on a number line is a _____.

10.   A number located to the right of 0 on a number line is a _____.

11.   Numbers that can be represented by point on the number line are _____.

**12.** _____ uses a variable and a description to describe a set.

**13.** A number that can be written as the quotient of two integers is a _____.

## Objective 1   Classify numbers and graph them on number lines.

*Use a real number to express each number in the following applications.*

**1.** Last year Nina lost 75 pounds.

**1.** _____

**2.** Mt. Whitney, one of the highest mountains in the United States, has an altitude of 14,495 feet.

**2.** _____

**3.** The Dead Sea, the saltiest body of water in the world, lies 396 meters below the level of the Mediterranean Sea.

**3.** _____

**4.** Between 1970 and 1982, the population of Norway increased by 279,867.

**4.** _____

*Graph each group of rational numbers on a number line.*

**5.** $-2, 3, -4, 1$

**5.**

```
<-+-+-+-+-+-+-+-+-+-+->
 -5-4-3-2-1 0 1 2 3 4 5
```

**6.** $\frac{1}{2}, 0, -3, -\frac{5}{2}$

**6.**

```
<-+-+-+-+-+-+-+-+-+-+->
 -5-4-3-2-1 0 1 2 3 4 5
```

**7.** $-3\frac{1}{2}, -\frac{3}{2}, 0, \frac{7}{2}, 1$

**7.**

```
<-+-+-+-+-+-+-+-+-+-+->
 -5-4-3-2-1 0 1 2 3 4 5
```

**8.** $-4.5, -2.3, 1.7, 4.2$

**8.**

```
<-+-+-+-+-+-+-+-+-+-+->
 -5-4-3-2-1 0 1 2 3 4 5
```

**Objective 2    Tell which of two real numbers is less than the other.**

*Select the smaller number in each pair.*

9.    $-5.99, -6.01$                                    9. _____

10.    $\frac{2}{3}, -\frac{1}{2}$                      10. _____

11.    $-(-4), -4$                                      11. _____

12.    $-\frac{2}{5}, -\frac{1}{4}$                     12. _____

*Decide whether each statement is **true** or **false**.*

13.    $-76 < 45$                                       13. _____

14.    $-5 > -5$                                        14. _____

15.    $-12 > -10$                                      15. _____

16.    $3 < -4$                                         16. _____

**Objective 3    Find the opposite of a real number.**

*Find the opposite of each number.*

17.    $-25$                                            17. _____

18.    $\frac{3}{8}$                                    18. _____

19.    $-(-22)$                                         19. _____

20.    $2\frac{3}{7}$                                   20. _____

21.    $0$                                              21. _____

22.    $4.5$                                            22. _____

**23.** $-\frac{5}{7}$                                      23. _____

**Objective 4    Find the absolute value of a real number.**

*Simplify.*

**24.** $-|95|$                                      24. _____

**25.** $-|49 - 39|$                                      25. _____

**26.** $\left|1\frac{1}{2} - 2\frac{1}{4}\right|$                                      26. _____

**27.** $\left|\frac{1}{2} + \frac{1}{3}\right|$                                      27. _____

**28.** $|-7.52 + 6.3|$                                      28. _____

**29.** $|16 - 14|$                                      29. _____

**30.** $|0|$                                      30. _____

# Chapter 9 THE REAL NUMBER SYSTEM

### 9.4     Adding Real Numbers

| Learning Objectives |
|---|
| 1     Add two numbers with the same sign. |
| 2     Add numbers with different signs. |
| 3     Add mentally. |
| 4     Use the rules for order of operations with real numbers. |
| 5     Translate words and phrases that indicate addition. |

### Key Terms

Use the vocabulary terms listed below to complete each statement in exercises 1–2.

       **sum**                **addends**

   **1.**   The answer to an addition problem is called the _____.

   **2.**   In an addition problem, the numbers being added are the _____.

### Objective 1    Add two numbers with the same sign.

*Find each sum.*

   **1.**    $-20 + (-20)$                                          **1.** _____

   **2.**    $9 + 12$                                                  **2.** _____

   **3.**    $-7 + (-11)$                                       **3.** _____

   **4.**    $-9 + (-9)$                                         **4.** _____

   **5.**    $\frac{3}{5} + \frac{4}{5}$                                            **5.** _____

   **6.**    $-2\frac{3}{8} + \left(-3\frac{1}{4}\right)$                              **6.** _____

## Objective 2    Add numbers with different signs.

*Use a number line to find each sum.*

**7.**    $7 + (-12)$

**7.** _____

**8.**    $-8 + 5$

**8.** _____

**9.**    $-4 + 4$

**9.** _____

*Find each sum.*

**10.**    $\dfrac{7}{12} + \left(-\dfrac{3}{4}\right)$

**10.** _____

**11.**    $-\dfrac{4}{7} + \dfrac{3}{5}$

**11.** _____

**12.**    $-10.475 + 6.325$

**12.** _____

## Objective 3    Add mentally.

*Perform each operation and then determine whether the statement is* true *or* false. *Try to do all work in your head.*

**13.**    $-14 + 11 = -3$

**13.** _____

**14.**    $(-14) + 15 + (-2) = -3$

**14.** _____

**15.**    $\dfrac{3}{5} + \left(-\dfrac{3}{10}\right) = -\dfrac{3}{10}$

**15.** _____

**16.**   $-\dfrac{3}{8}+\dfrac{11}{12}=-\dfrac{7}{24}$                16. _____

**17.**   $5\dfrac{3}{8}+\left(-4\dfrac{1}{2}\right)=2\dfrac{1}{8}$                17. _____

**18.**   $\left|-5+(-4)\right|=5+4$                18. _____

**Objective 4    Use the rules for order of operations with real numbers.**

*Find each sum.*

**19.**   $-14+3+\left[8+(-13)\right]$                19. _____

**20.**   $-2+\left[4+(-18+13)\right]$                20. _____

**21.**   $\left[(-7)+14\right]+\left[(-16)+3\right]$                21. _____

**22.**   $-8.9+\left[6.8+(-4.7)\right]$                22. _____

**23.**   $\dfrac{3}{8}+\left[-\dfrac{2}{3}+\left(-\dfrac{7}{12}\right)\right]$                23. _____

**24.**   $\left[-2\dfrac{3}{8}+\left(-3\dfrac{1}{4}\right)\right]+5\dfrac{3}{4}$                24. _____

## Objective 5    Translate words and phrases that indicate addition.

*Write a numerical expression for each phrase, and then simplify the expression.*

**25.**    The sum of −8 and −4 and −11

**25.** _____

_____

**26.**    The sum of −14 and −29, increased by 27

**26.** _____

_____

**27.**    −10 added to the sum of 20 and −4

**27.** _____

_____

*Solve each problem.*

**28.**    A football team gained 4 yards from scrimmage on the first play, lost 21 yards on the second play, and gained 9 yards on the third play. How many yards did the team gain or lose altogether? Write the answer as a signed number.

**28.** _____

**29.**    Pablo has $723 in his checking account. He write two checks, one for $358 and the other for $75. Finally, he deposits $205 in the account. How much does he now have in his account?

**29.** _____

**30.**    The temperature at dawn in Blackwood was 24°F. During the day the temperature decreased 30°. Then it increased 11° by sunset. What was the temperature at sunset?

**30.** _____

# Chapter 9 THE REAL NUMBER SYSTEM

### 9.5  Subtracting Real Numbers

| Learning Objectives |
| --- |
| 1  Find a difference. |
| 2  Use the definition of subtraction. |
| 3  Work subtraction problems that involve brackets. |
| 4  Translate words and phrases that indicate subtraction. |

### Key Terms

Use the vocabulary terms listed below to complete each statement in exercises 1–3.

**minuend**     **subtrahend**          **difference**

1.  The number from which another number is being subtracted is called the

   _____.

2.  The _____ is the number being subtracted.

3.  The answer to a subtraction problem is called the _____.

### Objective 1  Find a difference.

*Use a number line to find the difference.*

1.   $8 - 5$                                                           1. _____

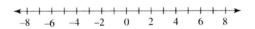

2.   $7 - 10$                                                          2. _____

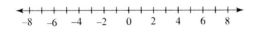

3.   $4 - 4$                                                           3. _____

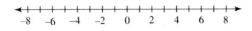

**4.**    $3 - 9$

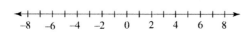

**4.** _____


**5.**    $-5 - 2$

**5.** _____


**6.**    $-3 - 5$

**6.** _____


## Objective 2    Use the definition of subtraction.

*Find each difference.*

**7.**    $-7 - (-14)$

**7.** _____


**8.**    $22 - (-24)$

**8.** _____


**9.**    $-5.6 - (-5.6)$

**9.** _____


**10.**    $-7.2 - 8.9$

**10.** _____


**11.**    $-3.2 - (-7.6)$

**11.** _____


**12.**    $\frac{1}{10} - \frac{1}{2}$

**12.** _____

**13.**    $-\dfrac{3}{10}-\left(-\dfrac{4}{15}\right)$                                 **13.** _____

**14.**    $3\dfrac{3}{4}-\left(-2\dfrac{1}{8}\right)$                                   **14.** _____

## Objective 3    Work subtraction problems that involve brackets.

*Perform each operation.*

**15.**    $\left[8-(-12)\right]-2$                               **15.** _____

**16.**    $-.2-\left[.6+(-.9)\right]$                            **16.** _____

**17.**    $\left[3-(-9)\right]-(-6)$                            **17.** _____

**18.**    $3-\left[-4+(11-19)\right]$                        **18.** _____

**19.**    $-2+\left[(-12+10)-(-4+2)\right]$          **19.** _____

**20.**    $\left(\dfrac{1}{2}-\dfrac{1}{3}\right)-\dfrac{5}{6}$                           **20.** _____

**21.**    $\dfrac{2}{9}-\left[\dfrac{5}{6}-\left(-\dfrac{2}{3}\right)\right]$                     **21.** _____

**22.** $\left[\frac{5}{8}-\left(-\frac{1}{16}\right)\right]-\left(-\frac{3}{8}\right)$                **22.** _____

## Objective 4    Translate words and phrases that indicate subtraction.

*Write a numerical expression for each phrase, and then simplify the expression.*

**23.**    4 less than −4                        **23.** _____

**24.**    −12 subtracted from the sum of −4 and −2      **24.** _____

**25.**    The sum of −4 and 12, decreased by 9        **25.** _____

**26.**    2 less than the difference between 10 and −4    **26.** _____

*Solve each problem.*

**27.**    Dr. Somers runs an experiment at −43.3°C. He then    **27.** _____
         lowers the temperature by 7.9°C. What is the new
         temperature for the experiment?

**28.**    David has a checking account balance of $439.42.    **28.** _____
         He overdraws his account by writing a check for
         $702.58. Write his new balance as a negative
         number.

**29.**    At 1:00 A.M., the temperature on the top of Mt.    **29.** _____
         Washington in New Hampshire was −12°F. At 11:00
         A.M., the temperature was 25°F. What was the rise
         in temperature?

**30.**    The highest point in a country has an elevation of    **30.** _____
         1408 meters. The lowest point is 396 meters below
         sea level. Using zero as sea level, find the difference
         between the two elevations.

# Chapter 9 THE REAL NUMBER SYSTEM

### 9.6  Multiplying and Dividing Real Numbers

| Learning Objectives |
|---|
| 1  Find the product of a positive number and a negative number. |
| 2  Find the product of two negative numbers. |
| 3  Use the reciprocal of a number to apply the definition of division. |
| 4  Use the rules for order of operations when multiplying and dividing signed numbers. |
| 5  Evaluate expressions involving variables. |
| 6  Translate words and phrases involving multiplication and division. |
| 7  Translate simple sentences into equations. |

### Key Terms

Use the vocabulary terms listed below to complete each statement in exercises 1–3.

**product**      **quotient**          **reciprocals**

1.  The answer to a division problem is called the _____.

2.  Pairs of numbers whose product is 1 are called _____.

3.  The answer to a multiplication problem is called the _____.

### Objective 1  Find the product of a positive number and a negative number.

*Find each product.*

1.  $7(-4)$                                    1. _____

2.  $\left(\frac{1}{5}\right)\left(-\frac{2}{3}\right)$                                    2. _____

3.  $\left(-\frac{3}{8}\right)\left(\frac{14}{9}\right)$                                    3. _____

4.  $(-3.2)(4.1)$                                    4. _____

## Objective 2    Find the product of two negative numbers.

*Find each product.*

**5.** $(-4)(-10)$                  **5.** _____

**6.** $\left(-\frac{2}{7}\right)\left(-\frac{14}{5}\right)$             **6.** _____

**7.** $(-1.3)(-2.1)$               **7.** _____

**8.** $(-.4)(-3.4)$                **8.** _____

## Objective 3    Use the reciprocal of a number to apply the definition of division.

*Find each quotient.*

**9.** $-\frac{3}{16} \div \frac{9}{8}$              **9.** _____

**10.** $\frac{-120}{-20}$                **10.** _____

**11.** $\frac{0}{-2}$                  **11.** _____

**12.** $\frac{10}{0}$                  **12.** _____

**13.** $5.5 \div (-2.2)$

**Objective 4** Use the rules for order of operations when multiplying and dividing signed numbers.

*Perform the indicated operations.*

**14.** $-4\big[(-2)(7)-2\big]$

14. _____

**15.** $-7\big[-4-(-2)(-3)\big]$

15. _____

**16.** $\dfrac{-7(2)-(-3)}{5+(-3)}$

16. _____

**17.** $\dfrac{-4\big[8-(-3+7)\big]}{-6\big[3-(-2)\big]-3(-3)}$

17. _____

**18.** $\dfrac{(-9+1)^2-(-6)(-2)}{5(-5)+3(4)}$

18. _____

**Objective 5** Evaluate expressions involving variables.

*Evaluate the following expressions if x = −3, y = 2, and a = 4.*

**19.** $-x+\big[(-a+y)-2x\big]$

19. _____

**20.** $(-4+x)(-a)-|x|$

20. _____

**21.**   $-x^2 + 2a^2 - 3y$                          **21.** _____

**22.**   $\dfrac{4a - x}{y^2}$                          **22.** _____

**Objective 6**   **Translate words and phrases involving multiplication and division.**

*Write a numerical expression for each phrase and simplify.*

**23.**   The product of –7 and 3, added to –7          **23.** _____

**24.**   Three-tenths of the difference between 50 and –10, subtracted from 85          **24.** _____

**25.**   The sum of –12 and the quotient of 49 and –7          **25.** _____

**26.**   The product of 40 and –3, divided by the difference between 5 and –10          **26.** _____

**Objective 7**   **Translate simple sentences into equations.**

*Write each statement in symbols, using x as the variable.*

**27.**   Two-thirds of a number is –7.          **27.** _____

**28.**   –8 times a number is 72.          **28.** _____

**29.**   When a number is divided by –4, the result is 1.          **29.** _____

**30.**   The quotient of –2 and a number is –9.          **30.** _____

# Chapter 9 THE REAL NUMBER SYSTEM

### 9.7    Properties of Real Numbers

| Learning Objectives | |
|---|---|
| 1 | Use the commutative properties. |
| 2 | Use the associative properties. |
| 3 | Use the identity properties. |
| 4 | Use the inverse properties. |
| 5 | Use the distributive property. |

### Key Terms

Use the vocabulary terms listed below to complete each statement in exercises 1–2.

**identity element for addition**

**identity element for multiplication**

1.    When the _____, 0, is added to a number, the number is unchanged.

2.    When a number is multiplied by the _____, 1, the number is unchanged.

### Objective 1    Use the commutative properties.

*Complete each statement. Use a commutative property.*

1.    $y + 4 = \underline{\hspace{1cm}} + y$                    1. _____

2.    $5(2) = \underline{\hspace{1cm}}(5)$                    2. _____

3.    $2 + \left[10 + (-9)\right] = \underline{\hspace{1cm}} + 2$                    3. _____

4.    $-4(4 + z) = \underline{\hspace{1cm}}(-4)$                    4. _____

5.    $10\left(\frac{1}{4} \cdot 2\right) = \underline{\hspace{1cm}}(10)$                    5. _____

**6.**     $3 \cdot (-2) + 12 = 12 + \underline{\hspace{1cm}}$        **6.** $\underline{\hspace{3cm}}$

## Objective 2    Use the associative properties.

*Complete each statement. Use an associative property.*

**7.**     $(4 \cdot 5)(-7) = \underline{\hspace{1cm}} \left[5(-7)\right]$        **7.** $\underline{\hspace{3cm}}$

**8.**     $\left[-4 + (-2)\right] + y = \underline{\hspace{1cm}} + (-2 + y)$        **8.** $\underline{\hspace{3cm}}$

**9.**     $4(ab) = \underline{\hspace{1cm}} \cdot b$        **9.** $\underline{\hspace{3cm}}$

**10.**     $\left[x + (-4)\right] + 3y = x + \underline{\hspace{1cm}}$        **10.** $\underline{\hspace{3cm}}$

**11.**     $(-r)\left[(-p)(-1)\right] = \underline{\hspace{1cm}} (-1)$        **11.** $\underline{\hspace{3cm}}$

**12.**     $4r + (3s + 14t) = \underline{\hspace{1cm}} + 14t$        **12.** $\underline{\hspace{3cm}}$

## Objective 3    Use the identity properties.

*Use an identity property to complete each statement.*

**13.**     $4 + 0 = \underline{\hspace{1cm}}$        **13.** $\underline{\hspace{3cm}}$

**14.**     $1(-4) = \underline{\hspace{1cm}}$        **14.** $\underline{\hspace{3cm}}$

**15.**     $\underline{\hspace{1cm}} \cdot 1 = 12$        **15.** $\underline{\hspace{3cm}}$

*Use an identity property to simplify each expression.*

**16.**   $\dfrac{30}{35}$

16. _____

**17.**   $\dfrac{7}{10} + \dfrac{9}{30}$

17. _____

**18.**   $\dfrac{27}{25} - \dfrac{8}{5}$

18. _____

## Objective 4   Use the inverse properties.

*Complete the statements so that they are examples of either an identity property or an inverse property. Identify which property is used.*

**19.**   $-4 + \underline{\hphantom{xxx}} = 0$

19. _____

**20.**   $\dfrac{2}{7} \cdot \underline{\hphantom{xxx}} = 1$

20. _____

**21.**   $-9 + \underline{\hphantom{xxx}} = -9$

21. _____

**22.**   $-\dfrac{3}{5} \cdot \underline{\hphantom{xxx}} = 1$

22. _____

**23.**   $\underline{\hphantom{xxx}} \cdot -2\dfrac{5}{6} = 1$

23. _____

**24.**   $\underline{\hphantom{xxx}} \cdot -2\dfrac{5}{6} = -2\dfrac{5}{6}$

24. _____

## Objective 5  Use the distributive property.

*Use the distributive property to rewrite each expression. Simplify if possible.*

**25.**  $-\left(-4b-8\right)$

**25.** _____

**26.**  $n\left(2a-4b+6c\right)$

**26.** _____

**27.**  $-2\left(5y-9z\right)$

**27.** _____

**28.**  $-\left(-2k+7\right)$

**28.** _____

**29.**  $-14x+\left(-14y\right)$

**29.** _____

**30.**  $2\left(7x\right)+2\left(8z\right)$

**30.** _____

# Chapter 9 THE REAL NUMBER SYSTEM

### 9.8    Simplifying Expressions

| **Learning Objectives** |
| --- |
| 1    Simplify expressions. |
| 2    Identify terms and numerical coefficients. |
| 3    Identify like terms. |
| 4    Combine like terms. |
| 5    Simplify expressions from word phrases. |

### Key Terms

Use the vocabulary terms listed below to complete each statement in exercises 1–3.

      **term**        **numerical coefficient**        **like terms**

1.  In the term $4x^2$, "4" is the_____.

2.  A number, a variable, or a product or quotient of a number and one or more variables raised to powers is called a _____.

3.  Terms with exactly the same variables, including the same exponents, are called _____.

### Objective 1    Simplify expressions.

*Simplify each expression.*

1.  $4(2x+5)+7$                                    1. _____

2.  $11-(d-2)+(-6)$                                2. _____

3.  $-4+s-(12-21)$                                 3. _____

4.  $-2(-5x+2)+7$                                  4. _____

5.  $7(5n-2)-(6-11)$                               5. _____

**6.** $4(-6p-2)+2-4$

**6.** _____

## Objective 2   Identify terms and numerical coefficients.

*Give the numerical coefficient of each term.*

**7.** $-2y^2$

**7.** _____

**8.** $125$

**8.** _____

**9.** $z^5$

**9.** _____

**10.** $-\dfrac{3}{5}a^2b$

**10.** _____

**11.** $\dfrac{7x}{9}$

**11.** _____

**12.** $5.6r^5$

**12.** _____

## Objective 3   Identify like terms.

*Identify each group of terms as **like** or **unlike**.*

**13.** $4x^2, -7x^2$

**13.** _____

**14.** $-8m, -8m^2$

**14.** _____

**15.** $7xy, -6xy^2$

**15.** _____

**16.** $2w, 4w, -w$

**16.** _____

**17.** $5z^3, 5z^2, 5z^2$

**17.** _____

**18.** $\dfrac{1}{3}, -\dfrac{3}{4}, 4$

**18.** _____

## Objective 4   Combine like terms.

*Simplify.*

**19.**   $7r - (2r + 4)$    **19.** _____

**20.**   $12y - 7y^2 + 4y - 3y^2$    **20.** _____

**21.**   $.8y^2 - .2xy - .3xy + .9y^2$    **21.** _____

**22.**   $-4(x + 4) + 2(3x + 1)$    **22.** _____

**23.**   $2.5(3y + 1) - 4.5(2y - 3)$    **23.** _____

**24.**   $\dfrac{1}{2}(2x - 4) - \dfrac{3}{4}(8x + 12)$    **24.** _____

## Objective 5   Simplify expressions from word phrases.

*Write each phrase as a mathematical expression and simplify by combining like terms. Use x as the variable.*

**25.**   The sum of six times a number and 12, added to four    **25.** _____
times the number.

**26.** The sum of seven times a number and 2, subtracted from three times the number.

**26.** _____

**27.** Three times the sum of 9 and twice a number, added to four times the number.

**27.** _____

**28.** The sum of ten times a number and 7, subtracted from the difference between 2 and nine times the number.

**28.** _____

**29.** Four times the difference between twice a number and six times the number, added to six times the sum of the number and 9.

**29.** _____

**30.** Four times the difference between twice a number and –10, subtracted from three times the sum of –7 and five times the number.

**30.** _____

# Chapter 10 EQUATIONS, INEQUALITIES, AND APPLICATIONS

## 10.1   The Addition Property of Equality

**Learning Objectives**
1   Identify linear equations.
2   Use the addition property of equality.
3   Simplify, and then use the addition property of equality.

## Key Terms

Use the vocabulary terms listed below to complete each statement in exercises 1–3.

**linear equation**      **solution set**      **equivalent equations**

1.   Equations that have exactly the same solutions sets are called
     _____.

2.   An equation that can be written in the form $Ax + B = C$, where $A$, $B$, and $C$ are real
     numbers and $A \neq 0$, is called a _____.

3.   The set of all numbers that satisfy an equation is called its
     _____.

## Objective 1   Identify linear equations.

*Tell whether each of the following is a linear equation.*

1.   $9x + 2 = 0$                                    1. _____

2.   $3x^2 + 4x + 3 = 0$                             2. _____

3.   $7x^2 = 10$                                     3. _____

4.   $3x^3 = 2x^2 + 5x$                              4. _____

5.   $\dfrac{5}{x} - \dfrac{3}{2} = 0$               5. _____

**6.**    $4x - 2 = 12x + 9$                          **6.** _____

## Objective 2    Use the addition property of equality.

*Solve each equation by using the addition property of equality. Check each solution.*

**7.**    $y - 4 = 16$                               **7.** _____

**8.**    $r + 9 = 8$                                **8.** _____

**9.**    $3x + 2 = 5x + 12$                         **9.** _____

**10.**   $3y = 7y - 4$                              **10.** _____

**11.**   $p - \frac{2}{3} = \frac{5}{6}$            **11.** _____

**12.**   $y + 4\frac{1}{2} = 3\frac{3}{4}$          **12.** _____

**13.**   $\frac{2}{3}t - 5 = \frac{5}{3}t$          **13.** _____

**14.**   $\frac{9}{8}p - \frac{1}{2} = \frac{1}{8}p$   **14.** _____

**15.**     $5.7x + 12.8 = 4.7x$                **15.** _____

**16.**     $9.5y - 2.4 = 10.5y$             **16.** _____

**17.**     $2z + 8 = -12$                    **17.** _____

**18.**     $-7t + 12 = -4t$                 **18.** _____

**Objective 3    Simplify, and then use the addition property of equality.**

*Solve each equation. First simplify each side of the equation as much as possible. Check each solution.*

**19.**     $3(t + 3) - (2t + 7) = 9$         **19.** _____

**20.**     $5x + 4(2x + 1) - (5x - 1 - 2) = 9$      **20.** _____

**21.**     $-4(5g - 7) + 3(8g - 3) = 15 - 4 + 3g$      **21.** _____

**22.**     $10x + 4x - 11x + 4 - 7 = 2 - 4x - 3 + 8x$      **22.** _____

**23.**     $4(3a - 2) - 6(2 + a) = 5(2a - 5)$         **23.** _____

**24.**     $2(4t + 6) - 3(2t - 3) = -3(3t - 4) + 5 - t$         **24.** _____

**25.**     $-7(1 + 2b) - 6(3 - 5b) = 5(4 + 3b) - 45$         **25.** _____

**26.**     $8(2 - 4b) + 3(5 - b) = 4(1 - 9b) + 22$         **26.** _____

**27.**     $\frac{8}{5}t + \frac{1}{3} = \frac{5}{6} + \frac{3}{5}t - \frac{1}{6}$         **27.** _____

**28.**     $\frac{5}{12} + \frac{7}{6}s - \frac{1}{6} = \frac{5}{6}s + \frac{1}{4} - \frac{2}{3}s$         **28.** _____

**29.**     $3.6p + 4.8 + 4.0p = 8.6p - 3.1 + .7$         **29.** _____

**30.**     $.03x + 0.6 + .09x - .9 = 2.1$         **30.** _____

# Chapter 10 EQUATIONS, INEQUALITIES, AND APPLICATIONS

### 10.2 The Multiplication Property of Equality

| **Learning Objectives** |
| --- |
| 1     Use the multiplication property of equality. |
| 2     Simplify, and then use the multiplication property of equality. |

### Key Terms

Use the vocabulary terms listed below to complete each statement in exercises 1–2.

    **multiplication property of equality**      **addition property of equality**

1.    The _____ states that multiplying both sides of an equation by the same nonzero number will not change the soltuion.

2.    When the same quantity is added to both sides of an equation, the _____ is being applied.

### Objective 1    Use the multiplication property of equality.

*Solve each equation and check your solution.*

1.    $8x = 24$                                     1. _____

2.    $-3w = 51$                                  2. _____

3.    $-16a = -48$                               3. _____

4.    $\dfrac{b}{5} = 4$                                     4. _____

5.    $\dfrac{3p}{7} = -6$                                 5. _____

**6.**  $\dfrac{b}{-2} = 21$

**6.** _____

**7.**  $-\dfrac{7}{2}t = -4$

**7.** _____

**8.**  $\dfrac{y}{4} = \dfrac{1}{3}$

**8.** _____

**9.**  $\dfrac{6}{7}y = \dfrac{2}{3}$

**9.** _____

**10.**  $\dfrac{3}{4}r = -27$

**10.** _____

**11.**  $.81m = 2.916$

**11.** _____

**12.**  $2.1a = 9.03$

**12.** _____

**13.**  $7.5p = -61.5$

**13.** _____

**14.**     $-2.7v = -17.28$

**14.** _____

**15.**     $4.3r = -11.61$

**15.** _____

**Objective 2**     **Simplify, and then use the multiplication property of equality.**

*Solve each equation and check your solution.*

**16.**     $12r + 3r = -90$

**16.** _____

**17.**     $-7b + 12b = 125$

**17.** _____

**18.**     $3b - 4b = 8$

**18.** _____

**19.**     $7q - 10q = -24$

**19.** _____

**20.**     $3w - 7w = 20$

**20.** _____

**21.**     $10a - 7a = -24$

**21.** _____

**22.**     $6m - 14m = -56$

**22.** _____

**23.**    $.8c + .6c = -2.1$                          **23.** _____

**24.**    $8f + 4f - 3f = 108$                         **24.** _____

**25.**    $8s - 3s + 4s = 90$                          **25.** _____

**26.**    $4x - 8x + 2x = 18$                          **26.** _____

**27.**    $2f + 3f - 7f = 48$                          **27.** _____

**28.**    $-11h - 6h + 14h = -21$                      **28.** _____

**29.**    $18r - 6r + 3r = -105$                       **29.** _____

**30.**    $17x + 9x - 11x = -9$                        **30.** _____

# Chapter 10 EQUATIONS, INEQUALITIES, AND APPLICATIONS

### 10.3    More on Solving Linear Equations

| Learning Objectives |
| --- |
| 1        Learn and use the four steps for solving a linear equation. |
| 2        Solve equations with fractions or decimals as coefficients. |
| 3        Solve equations that have no solution or infinitely many solutions. |
| 4        Write expressions for two related unknown quantities. |

**Key Terms**

Use the vocabulary terms listed below to complete each statement in exercises 1–3.

**conditional equation          identity                    contradiction**

1.    An equation with no solution is called a(n) _____.

2.    A(n) _____ is an equation that is true for
      some values of the variable and false for other values.

3.    An equation that is true for all values of the variable is called a(n)
      _____.

### Objective 1    Learn and use the four steps for solving a linear equation.

*Solve each equation and check your solution.*

1.    $7t + 6 = 11t - 4$                          1. _____

2.    $4(z - 2) - (3z - 1) = 2z - 6$              2. _____

3.    $3(x + 4) = 6 - 2(x - 8)$                    3. _____

4.    $-(v + 2) = 3 + v$                           4. _____

**5.**  $3 - (1 - y) = 3 + 5y$

**6.**  $3a - 6a + 4(a - 4) = -2(a + 2)$

**7.**  $3(t + 5) = 6 - 2(t - 4)$

**8.**  $4r - 3(3r - 2) = 8 - 3(r - 4)$

**Objective 2    Solve equations with fractions or decimals as coefficients.**

*Solve each equation and check your solution.*

**9.**  $\dfrac{3}{8}x - \dfrac{1}{3}x = \dfrac{1}{12}$

**10.**  $\dfrac{1}{3}(2m - 1) - \dfrac{3}{4}m = \dfrac{5}{6}$

**11.**  $\dfrac{5}{6}(r - 2) - \dfrac{2}{9}(r + 4) = \dfrac{7}{18}$

**12.**  $\dfrac{3}{8}x - \left(x - \dfrac{3}{4}\right) = \dfrac{5}{8}(x + 3)$

**13.**    $.90x = .40(30) + .15(100)$                         **13.** _____

**14.**    $.35(20) + .45y = .125(200)$                        **14.** _____

**15.**    $.24x - .38(x + 2) = -.34(x + 4)$                   **15.** _____

**16.**    $.45a - .35(20 - a) = .02(50)$                      **16.** _____

**Objective 3     Solve equations that have no solution or infinitely many solutions.**

*Solve each equation and check your solution.*

**17.**    $3(6x - 7) = 2(9x - 6)$                             **17.** _____

**18.**    $6y - 3(y + 2) = 3(y - 2)$                          **18.** _____

**19.**    $-1 - (2 + y) = -(-4 + y)$                          **19.** _____

**20.**    $6(6t + 1) = 9(4t - 3) + 33$                        **20.** _____

**21.** $3(r-2)-r+4=2r+6$        **21.** _____

**22.** $8(2d-4)-3(7d+8)=-5(d+2)$      **22.** _____

**23.** $2(5w-3)-6=3(3w+1)+5(w-3)-4w$    **23.** _____

**Objective 4    Write expressions for two related unknown quantities.**

*Write an expression for the two related unknown quantities.*

**24.** Two numbers have a sum of 36. One is *m*. Find the    **24.** _____
other number.

**25.** The product of two numbers is 17. One number is *p*.   **25.** _____
What is the other number?

**26.** A cashier has *q* dimes. Find the value of the dimes in   **26.** _____
cents.

**27.** The length of a rectangle is *x* inches. Its width is four   **27.** _____
times the length. Find the width of the rectangle.

**28.** Temperature in degrees Kelvin is always 273 more    **28.** _____
than temperature in degrees Celsius. If the Kelvin
temperature is $k°$, what is the temperature in degree
Celsius?

**29.** Shirley is *x* years old. Her mother is 28 years older.    **29.** _____
How old is her mother?

**30.** Admission to the circus costs *x* dollars for an adult    **30.** _____
and *y* dollars for a child. Find the total cost of 6
adults and 4 children.

# Chapter 10 EQUATIONS, INEQUALITIES, AND APPLICATIONS

### 10.4  An Introduction to Applications of Linear Equations

| Learning Objectives |
|---|
| 1    Learn the six steps for solving applied problems. |
| 2    Solve problems involving unknown numbers. |
| 3    Solve problems involving sums of quantities. |
| 4    Solve problems involving supplementary and complementary angles. |
| 5    Solve problems involving consecutive integers. |

### Key Terms

Use the vocabulary terms listed below to complete each statement in exercises 1–5.

> **complementary angles**     **right angle**     **supplementary angles**
>
> **straight angle**        **consecutive integers**

1. Two angles whose measures sum to 180° are _____.

2. Two angles whose measures sum to 90° are _____.

3. An angle whose measure is exactly 90° is a _____.

4. An angle whose measure is exactly 180° is a _____.

5. Two integers that differ by 1 are _____.

### Objective 1    Learn the six steps for solving applied problems.

1. Write the six problem-solving steps.          1. _____

_____

_____

_____

_____

_____

**Objective 2   Solve problems involving unknown numbers.**

*Write an equation for each of the following and then solve the problem. Use x as the variable.*

2.   If 4 is added to 3 times a number, the result is 7. Find the number.

2.   _____

_____

3.   If 2 is subtracted from four times a number, the result is 3 more than six times the number. What is the number?

3.   _____

_____

4.   If –2 is multiplied by the difference between 4 and a number, the result is 24. Find the number.

4.   _____

_____

5.   Six times the difference between a number and 4 equals the product of the number and –2. Find the number.

5.   _____

_____

6.   When the difference between a number and 4 is multiplied by –3, the result is two more than –5 times the number. Find the number.

6.   _____

_____

7.   If four times a number is added to 7, the result is five less than six times the number. Find the number.

7.   _____

_____

**Objective 3    Solve problems involving sums of quantities.**

*Write an equation for each of the following and then solve the problem. Use x as the variable.*

8.   A rope 116 inches long is cut into three pieces. The middle-sized piece is 10 inches shorter than twice the shortest piece. The longest piece is $\frac{5}{3}$ as long as the shortest piece. What is the length of the shortest piece?

8. _____

_____

9.   George and Al were opposing candidates in the school board election. George received 21 more votes than Al, with 439 votes cast. How many votes did Al receive?

9. _____

_____

10.   On a psychology test, the highest grade was 38 points more than the lowest grade. The sum of the two grades was 142. Find the lowest grade.

10. _____

_____

11.   Mount McKinley is Alaska is 5910 feet higher than Mount Rainier in Washington. Together, their heights total 34,730 feet. How high is each mountain?

11. _____

Mt. Rainier _____

Mt. McKinley _____

12.   Charles bought five general admission tickets and four student tickets for a movie. He paid $35.25. If each student ticket cost $3.50, how much did each general admission ticket cost?

12. _____

_____

**13.** Penny is making punch for a party. The recipe requires twice as much orange juice as cranberry juice and 8 times as much ginger ale as cranberry juice. If she plans to make 176 ounces of punch, how much of each ingredient should she use?

13. _____

cranberry juice _____

orange juice _____

ginger ale _____

**14.** Pablo, Faustino, and Mark swim at a public pool each day for exercise. One day Pablo swam five more than three times as many laps as Mark, and Faustino swam four times as many laps as Mark. If the men swam 29 laps altogether, how many laps did each one swim?

14. _____

Mark _____

Pablo_____

Faustino _____

**15.** Linda wishes to build a rectangular dog pen using 52 feet of fence and the back of her house, which is 36 feet long to enclose the pen. How wide will the dog pen be if the pen is 36 feet long?

15. _____

_____

**Objective 4    Solve problems involving supplementary and complementary angles.**

*Solve each problem.*

**16.** Find the measure of an angle if the measure of the angle is 8° less than three times the measure of its supplement.

16. _____

**17.** Find the measure of an angle whose supplement measures 20° more than twice its complement.

17. _____

**18.** Find the measure of an angle such that the sum of the measures of its complement and its supplement is 138°.

**18.** _____

**19.** Find the measure of an angle such that the difference between the measure of its supplement and twice the measure of its complement is 49°.

**19.** _____

**20.** Find the measure of an angle whose complement is 9° more than twice its measure.

**20.** _____

**21.** Find the measure of an angle such that the difference between the measures of an angle and its complement is 20°.

**21.** _____

**22.** Find the measure of an angle if its supplement measures 4° less than three times its complement.

**22.** _____

**Objective 5    Solve problems involving consecutive integers.**

_Solve each problem._

**23.** Find two consecutive even integers whose sum is 154.

**23.** _____

**24.** Find two consecutive even integers such that the smaller, added to twice the larger, is 292.

**24.** _____

**25.** Find two consecutive integers such that the larger, added to three times the smaller, is 109.

**25.** _____

**26.** Find two consecutive odd integers such that if three times the smaller is added to twice the larger, the sum is 69.

**26.** _____

**27.** Find two consecutive odd integers such that the larger, added to eight times the smaller, equals 119.

**27.** _____

**28.** Find three consecutive odd integers whose sum is 363.

**28.** _____

**29.** Find three consecutive integers such that the sum of the first two is 74 more than the third.

**29.** _____

**30.** The sum of four consecutive even integers is 4. Find the integers.

**30.** _____

# Chapter 10 EQUATIONS, INEQUALITIES, AND APPLICATIONS

## 10.5    Formulas and Additional Applications from Geometry

| **Learning Objectives** |
| :--- |
| 1    Solve a formula for one variable, given the values of the other variables. |
| 2    Use a formula to solve an applied problem. |
| 3    Solve problems involving vertical angles and straight angles. |
| 4    Solve a formula for a specified variable. |

### Key Terms

Use the vocabulary terms listed below to complete each statement in exercises 1–4.

**formula        area            perimeter    vertical angles**

1.    The nonadjacent angles formed by two intersecting lines are called

_____.

2.    An equation in which variables are used to describe a relationship is called a(n)

_____.

3.    The distance around a figure is called its _____.

4.    A measure of the surface covered by a figure is called its _____.

### Objective 1    Solve a formula for one variable, given the values of the other variables.

*In the following exercises, a formula is given, along with the values of all but one of the variables in the formula. Find the value of the variable that is not given.*

1.    $V = LWH$; $L = 2, W = 4, H = 3$          1. _____

2.    $P = 2L + 2W$; $P = 42$; $W = 6$          2. _____

3.    $S = \dfrac{a}{1-r}$; $S = 60, r = .4$          3. _____

**4.**     $I = prt$; $I = 288$, $r = .04$, $t = 3$                    4. _____

**5.**     $C = \frac{5}{9}(F - 32)$; $F = 104$                        5. _____

**6.**     $A = \frac{1}{2}(b + B)h$; $b = 6$, $B = 16$, $A = 132$     6. _____

**7.**     $V = \frac{1}{3}\pi r^2 h$; $r = 4$, $h = 6$, $\pi = 3.14$   7. _____

**Objective 2    Use a formula to solve an applied problem.**

*Use a formula to write an equation for each of the following applications; then solve the application. (Use 3.14 as an approximation for π.)*

**8.**     Find the length of a rectangular garden if its        8. _____
           perimeter is 96 feet and its width is 12 feet.

**9.**     Find the height of a triangular banner whose area is   9. _____
           48 square inches and base is 12 inches.

**10.** Ruth has 42 feet of binding for a rectangular rug that she is weaving. If the rug is 9 feet wide, how long can she make the rug if she wishes to use all the binding on the perimeter of the rug?

10. _____

**11.** Linda invests $5000 at 6% simple interest and earns $450. How long did Linda invest her money?

11. _____

**12.** A tent has the shape of a right pyramid. The volume is 200 cubic feet and the height is 12 feet. Find the area of the floor of the tent.

12. _____

**13.** A spherical balloon has a radius of 9 centimeters. Find the amount of air required to fill the balloon. (Round your answer to the nearest hundredth.)

13. _____

**14.** Find the height of an ice cream cone if the diameter is 6 centimeters and the volume is 37.68 cubic centimeters. (Round your answer to the nearest hundredth.)

14. _____

**15.** The circumference of a circular garden is 628 feet. Find the area of the garden. (Hint: First find the radius of the garden.)

15. _____

Name:                           Date:

Instructor:                 Section:

## Objective 3     Solve problems involving vertical angles and straight angles.

*Find the measure of each marked angle.*

**16.**

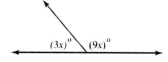

**16.** _____

**17.**

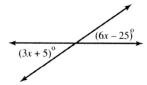

**17.** _____

**18.**

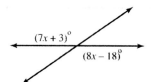

**18.** _____

**19.**

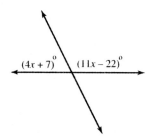

**19.** _____

**20.**

**20.** _____

**21.**

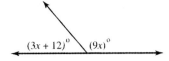

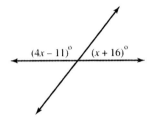

21. _____

**22.**

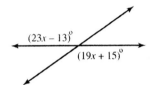

22. _____

**23.**

23. _____

## Objective 4    Solve a formula for a specified variable.

*Solve each formula for the specified variable.*

**24.**    $V = LWH$  for $H$

24. _____

**25.**    $S = \dfrac{a}{1-r}$  for $r$

25. _____

**26.**   $a_n = a_1 + (n-1)d$  for $n$                          **26.** _____

**27.**   $P = A - Art$  for $A$                                **27.** _____

**28.**   $S_n = \frac{n}{2}(a_1 + a_n)$  for $a_1$             **28.** _____

**29.**   $S = (n-2)180$  for $n$                               **29.** _____

**30.**   $V = \frac{1}{3}\pi r^2 h$  for $h$                   **30.** _____

# Chapter 10 EQUATIONS, INEQUALITIES, AND APPLICATIONS

### 10.6 Solving Linear Inequalities

| **Learning Objectives** |
| --- |
| 1      Graph intervals on a number line. |
| 2      Use the addition property of inequality. |
| 3      Use the multiplication property of inequality. |
| 4      Solve inequalities using both properties of inequality. |
| 5      Use inequalities to solve applied problems. |
| 6      Solve linear inequalities with three parts. |

### Key Terms

Use the vocabulary terms listed below to complete each statement in exercises 1–5.

> **inequalities    interval      interval notation      linear inequality**
>
> **three-part inequality**

1. An inequality that says that one number is between two other numbers is a(n)_____.

2. A portion of a number line is called a(n) _____.

3. A(n) _____ can be written in the form $Ax + B > C$, $Ax + B \geq C$, $Ax + B < C$, or $Ax + B \leq C$, where $A$, $B$, and $C$ are real numbers with $A \neq 0$.

4. Algebraic expressions related by $>$, $\geq$, $<$, or $\leq$ are called _____.

5. The _____ for $a \leq x < b$ is $[a, b)$.

### Objective 1    Graph intervals on a number line.

*Write each inequality in interval notation and graph the interval.*

1.   $3 < a$

     1. _____

2.   $y \geq -2$

     2. _____

3.   $-1 < x < 3$

     3. _____

**4.**   $-3 \le y < 0$

4. _____

**5.**   $-3 < a \le 2$

5. _____

## Objective 2   Use the addition property of inequality.

*Solve each inequality. Write the solution set in interval notation and then graph it.*

**6.**   $5a + 3 \le 6a$

6. _____

**7.**   $-2 + 8b \ge 7b - 1$

7. _____

**8.**   $6 + 3x < 4x + 4$

8. _____

**9.**   $3 + 5p \le 4p + 3$

9. _____

**10.**   $9 + 8b > 9b + 11$

10. _____

## Objective 3   Use the multiplication property of inequality.

*Solve each inequality. Write the solution set in interval notation and then graph it.*

**11.**   $-2s < 4$

11. _____

**12.**   $4k \geq -16$

**12.** _____

**13.**   $\frac{3}{5}n \geq 0$

**13.** _____

**14.**   $-5t \leq -35$

**14.** _____

**15.**   $-9m > -36$

**15.** _____

## Objective 4   Solve inequalities using both properties of inequality.

*Solve each inequality. Write the solution set in interval notation and then graph it.*

**16.**   $4(y-3)+2 > 3(y-2)$

**16.** _____

**17.**   $-3(m+4)+1 \leq -4(m-2)$

**17.** _____

**18.**   $7(2-x)-3 \leq -2(x-4)-x$

**18.** _____

19.  $3x - \frac{3}{4} \geq 2x + \frac{1}{3}$

19. _____

20.  $5(2z + 2) - 2(z - 3) > 3(2z + 5) + z$

20. _____

**Objective 5   Use inequalities to solve applied problems.**

*Solve each problem.*

21.  Lauren has grades of 98 and 86 on her first two chemistry quizzes. What must she score on her third quiz to have an average of at least 91 on the three quizzes?

21. _____

22.  Nina has a budget of $230 for gifts for this year. So far she has bought gifts costing $47.52, $38.98, and $26.98. If she has three more gifts to buy, find the average amount she can spend on each gift and still stay within her budget.

22. _____

**23.**    Ruth tutors mathematics in the evenings in an office          **23.** _____
for which she pays $600 per month rent. If rent is
her only expense and she charges each student $40
per month, how many students must she teach to
make a profit of at least $1600 per month?

**24.**    Two sides of a triangle are equal in length, with the          **24.** _____
third side 8 feet longer than one of the equal sides.
The perimeter of the triangle cannot be more than 38
feet. Find the largest possible value for the length of
the equal sides.

**25.**    If twice the sum of a number and 7 is subtracted             **25.** _____
from three times the number, the result is more
than –9. Find all such numbers.

**Objective 6    Solve linear inequalities with three parts.**

*Solve each inequality. Write the solution set in interval notation and then graph it.*

**26.**    $9 < 2x + 1 \le 15$                                        **26.** _____

**27.** $\quad -17 \leq 3x - 2 < -11$

**27.** _____

&lt;+++++++++++++++++&gt;

**28.** $\quad 6 < 2x - 4 < 8$

**28.** _____

&lt;+++++++++++++++++&gt;

**29.** $\quad -5 < -2 - x \leq 4$

**29.** _____

&lt;+++++++++++++++++&gt;

**30.** $\quad 1 < 3z + 4 < 19$

**30.** _____

&lt;+++++++++++++++++&gt;

# Chapter 11 GRAPHS OF LINEAR EQUATIONS AND INEQUALITIES IN TWO VARIABLES

## 11.1    Reading Graphs; Linear Equations in Two Variables

| Learning Objectives |  |
| --- | --- |
| 1 | Interpret graphs. |
| 2 | Write a solution as an ordered pair. |
| 3 | Decide whether a given ordered pair is a solution of a given equation. |
| 4 | Complete ordered pairs for a given equation. |
| 5 | Complete a table of values. |
| 6 | Plot ordered pairs. |

## Key Terms

Use the vocabulary terms listed below to complete each statement in exercises 1–14.

> **bar graph     line graph     linear equation in two variables**
>
> **ordered pair            table of values        *x*-axis**
>
> **_y_-axis      rectangular (Cartesian) coordinate system**
>
> **quadrants   origin        plane                 coordinates**
>
> **plot        scatter diagram**

1.    A _____ uses dots connected by line to show trends.

2.    An equation that can be written in the form $Ax + By = C$, where $A$, $B$, and $C$ are real numbers and $A$, $B \neq 0$, is called a _____.

3.    A _____ uses bars of various heights or lengths to show quantity or frequency.

4.    _____ are the numbers in the ordered pair that specify the location of a point on a rectangular coordinate system.

5.    In a rectangular coordinate system, the horizontal axis is called the _____.

6.    In a rectangular coordinate system, the vertical axis is called the _____.

7.    A pair of numbers written between parentheses in which order is important is called a(n) _____.

8.    Together, the *x*-axis and the *y*-axis form a _____.

9.    A coordinate system divides the plane into four regions called _____.

**10.** The axis lines in a coordinate system intersect at the _____.

**11.** To _____ an ordered pair is to find the corresponding point on a coordinate system.

**12.** A graph of ordered pairs is called a _____.

**13.** A table showing selected ordered pairs of numbers that satisfy an equation is called a _____.

**14.** A flat surface determined by two intersecting lines is a _____.

## Objective 1   Interpret graphs.

*The bar graph shows the enrollment for the fall semester at a small college for the past five years. Use this graph for problems 1–2.*

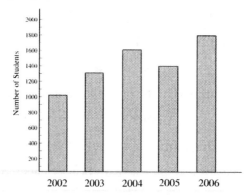

**1.** In what years was enrollment less than 1500 students?

**1.** _____

**2.** How many more students were enrolled in 2004 than in 2003?

**2.** _____

*The line graph shows the number of degrees awarded by a university for the years 2000–2005. Use this graph to answer exercises 3–5.*

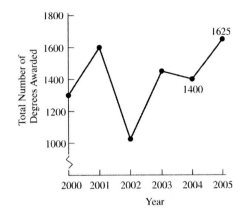

3. Between which two years did the total number of degrees awarded show the smallest change?

3. _____

4. Between which two years did the total number of degrees awarded show the greatest decline?

4. _____

5. If 20% of the degrees awarded in 2005 were M.B.A. degrees, how many M.B.A.'s were awarded in 2005?

5. _____

## Objective 2  Write a solution as an ordered pair.

*Write each solution as an ordered pair.*

6. $x = 4$ and $y = 7$

6. _____

7. $x = \frac{1}{3}$ and $y = -9$

7. _____

8. $y = \frac{1}{3}$ and $x = 0$

8. _____

9. $x = .2$ and $y = .3$

9. _____

**10.** $\quad x = -4$ and $y = 0$                  **10.** _____

**Objective 3    Decide whether a given ordered pair is a solution of a given equation.**

*Decide whether the given ordered pair is a solution of the given equation.*

**11.** $\quad 4x - 3y = 10;\ (1, 2)$                **11.** _____

**12.** $\quad 2x - 3y = 1;\ \left(0, \frac{1}{3}\right)$           **12.** _____

**13.** $\quad x = -7;\ (-7, 9)$               **13.** _____

**14.** $\quad x = 1 - 2y;\ \left(0, -\frac{1}{2}\right)$         **14.** _____

**15.** $\quad 2x = 3y;\ (3, 2)$               **15.** _____

## Objective 4    Complete ordered pairs for a given equation.

*For each of the given equations, complete the ordered pairs beneath it.*

**16.**    $y = 2x - 5$

    (a)    $(2, \phantom{0})$

    (b)    $(0, \phantom{0})$

    (c)    $(\phantom{0}, 3)$

    (d)    $(\phantom{0}, -7)$

    (e)    $(\phantom{0}, 9)$

**16.**

(a) _____

(b) _____

(c) _____

(d) _____

(e) _____

**17.**    $y = 3 + 2x$

    (a)    $(-4, \phantom{0})$

    (b)    $(2, \phantom{0})$

    (c)    $(\phantom{0}, 0)$

    (d)    $(-2, \phantom{0})$

    (e)    $(\phantom{0}, -7)$

**17.**

(a) _____

(b) _____

(c) _____

(d) _____

(e) _____

**18.**    $x = -2$

    (a)    $(\phantom{0}, -2)$

    (b)    $(\phantom{0}, 0)$

    (c)    $(\phantom{0}, 19)$

    (d)    $(\phantom{0}, 3)$

    (e)    $\left(\phantom{0}, -\frac{2}{3}\right)$

**18.**

(a) _____

(b) _____

(c) _____

(d) _____

(e) _____

**19.**    $y = 4$

    (a)    $(2, \phantom{0})$

    (b)    $(0, \phantom{0})$

    (c)    $(4, \phantom{0})$

    (d)    $(-4, \phantom{0})$

    (e)    $(.75, \phantom{0})$

**19.**

(a) _____

(b) _____

(c) _____

(d) _____

(e) _____

**20.**  $5x + 4y = 10$

    (a)  $(2, \ )$

    (b)  $(4, \ )$

    (c)  $(\ , 3)$

    (d)  $(0, \ )$

    (e)  $(\ , 2)$

**20.**

    (a)  _____

    (b)  _____

    (c)  _____

    (d)  _____

    (e)  _____

## Objective 5    Complete a table of values.

*Complete each table of values. Write the results as ordered pairs.*

**21.**  $3x - 4y = -6$

| $x$ | $y$ |
|---|---|
| 0 |   |
|   | 0 |
| 2 |   |

**21.** _____

**22.**  $-7x + 2y = -14$

| $x$ | $y$ |
|---|---|
|   | 0 |
| 0 |   |
| 3 |   |

**22.** _____

**23.**  $2x + 5 = 7$

| $x$ | $y$ |
|---|---|
|   | -3 |
|   | 0 |
|   | 5 |

**23.** _____

**24.**   $y - 4 = 0$

| $x$ | $y$ |
|-----|-----|
| $-4$ |    |
| $0$ |    |
| $6$ |    |

**24.** _____

**25.**   $4x + 3y = 12$

| $x$ | $y$ |
|-----|-----|
| $0$ |    |
|     | $0$ |
|     | $-1$ |

**25.** _____

## Objective 6    Plot ordered pairs.

*Plot the each ordered pair on a coordinate system.*

**26.**   $(0, -2)$

**27.**   $(2, 5)$

**28.**   $(-2, -7)$

**29.**   $(-3, 4)$

**30.**   $(4, -4)$

**26.–30.**

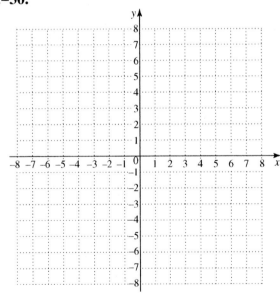

Name:                                           Date:
Instructor:                                     Section:

# Chapter 11 GRAPHS OF LINEAR EQUATIONS AND INEQUALITIES IN TWO VARIABLES

## 11.2    Graphing Linear Equations in Two Variables

| Learning Objectives |
|---|
| **1**    Graph linear equations by plotting ordered pairs. |
| **2**    Find intercepts. |
| **3**    Graph linear equations of the form $Ax + By = 0$. |
| **4**    Graph linear equations of the form $y = k$ or $x = k$. |
| **5**    Use a linear equation to model data. |

### Key Terms

Use the vocabulary terms listed below to complete each statement in exercises 1–4.

**graph          graphing      *y*-intercept     *x*-intercept**

1.    If a graph intersects the *y*-axis at *k*, then the _____ is $(0, k)$.

2.    If a graph intersects the *x*-axis at *k*, then the _____ is $(k, 0)$.

3.    The process of plotting the ordered pairs that satisfy a linear equation and drawing a line through them is called _____.

4.    The set of all points that correspond to the ordered pairs that satisfy the equation is called the _____ of the equation.

### Objective 1    Graph linear equations by plotting ordered pairs.

*Complete the ordered pairs for each equation. Then graph the equation by plotting the points and drawing a line through them.*

1.    $x + y = 3$            1.

     $(0,\ \ )$

     $(\ \ ,0)$

     $(2,\ \ )$

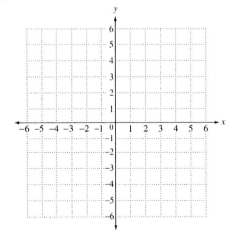

**2.**   $y = 3x - 2$

(0,  )

(  ,0)

(2,  )

**2.**

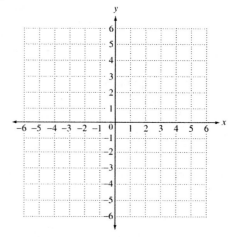

**3.**   $x - y = 4$

(0,  )

(  ,0)

(−2,  )

**3.**

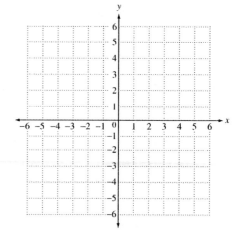

**4.**   $2y - 4 = x$

(0,  )

(  ,0)

(−2,  )

**4.**

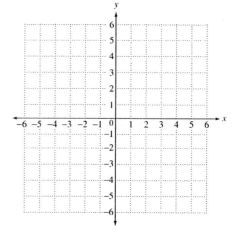

**5.**     $2x + 3y = 6$

$(0, \ \ )$

$(\ \ , 0)$

$(-3, \ \ )$

**5.**

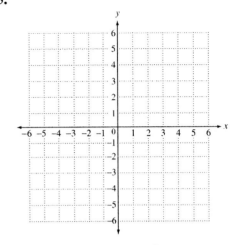

**6.**     $x = 2y + 1$

$(0, \ \ )$

$(\ \ , 0)$

$(\ \ , -2)$

**6.**

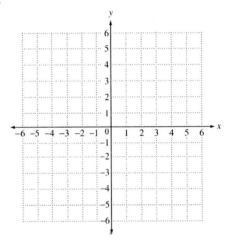

## Objective 2    Find intercepts.

*Find the intercepts for each equation. Then graph the equation.*

**7.**     $3x + y = 6$

**7.**

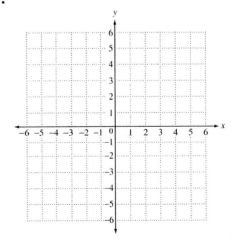

**8.**    $4x - y = 4$

**8.**

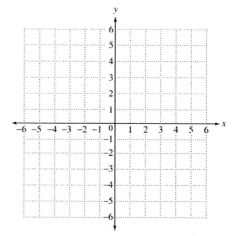

**9.**    $5x - 2y = -10$

**9.**

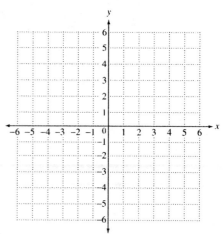

**10.**    $2x - 3y = 6$

**10.**

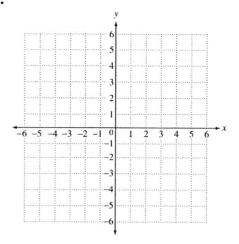

**11.** $4x - 7y = -8$                   **11.**

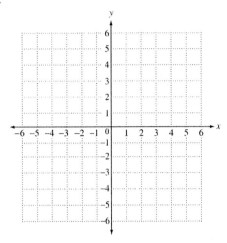

**12.** $3x - 2y = 8$                    **12.**

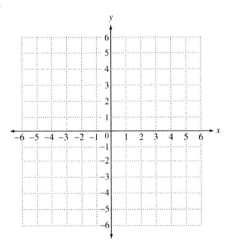

**Objective 3     Graph linear equations of the form $Ax + By = 0$.**

*Graph each equation.*

**13.** $-3x - 2y = 0$                   **13.**

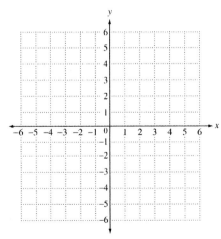

**14.**    $3x - y = 0$                          **14.**

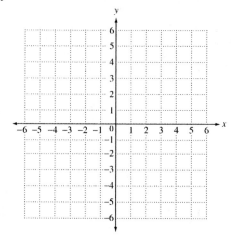

**15.**    $x + 5y = 0$                          **15.**

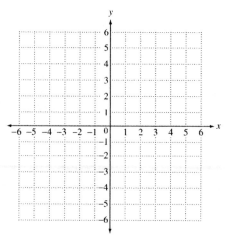

**16.**    $x + y = 0$                           **16.**

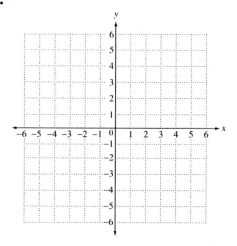

**17.**     $4x = 3y$

**17.**

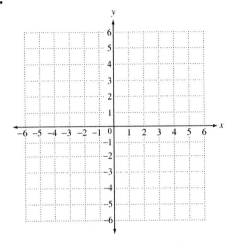

**18.**     $y = 2x$

**18.**

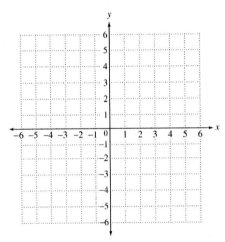

**Objective 4     Graph linear equations of the form $y = k$ or $x = k$.**

*Graph each equation.*

**19.**     $y = -2$

**19.**

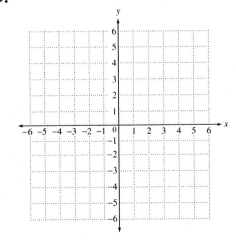

**20.** $x + 4 = 0$

**20.**

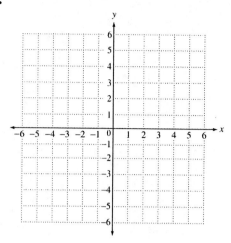

**21.** $x - 1 = 0$

**21.**

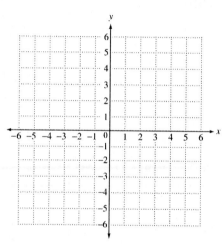

**22.** $y - 4 = 0$

**22.**

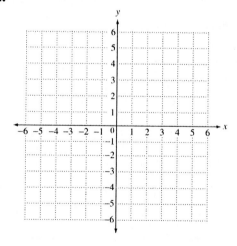

**23.**     $y + 3 = 0$

**23.**

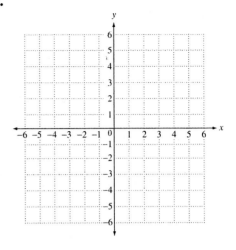

**24.**     $x = 0$

**24.**

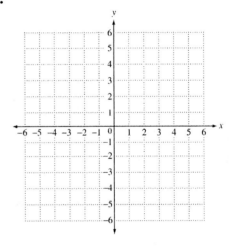

**Objective 5    Use a linear equation to model data.**

*Solve each problem. Then graph the equation.*

25.    The enrollment at Lincolnwood High School
       decreased during the years 2000 to 2005. If $x = 0$
       represents 2000, $x = 1$ represents 2001, and so on,
       the number of students enrolled in the school can be
       approximated by the equation $y = -85x + 2435$. Use
       this equation to approximate the number of students
       in each year from 2000 through 2005.

25. 2000 _____

    2001 _____

    2002 _____

    2003 _____

    2004 _____

    2005 _____

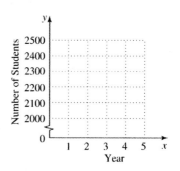

26.    The profit $y$ in millions of dollars earned by a small
       computer company can be approximated by the
       linear equation $y = .63x + 4.9$, where $x = 0$
       corresponds to 2004, $x = 1$ corresponds to 2005, and
       so on. Use this equation to approximate the profit in
       each year from 2004 through 2007.

26. 2004 _____

    2005 _____

    2006 _____

    2007 _____

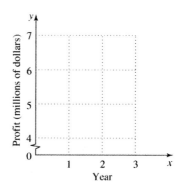

**27.** The number of band instruments sold by Elmer's Music Shop can be approximated by the equation $y = 325 + 42x$, where $y$ is the number of instruments sold and $x$ is the time in years, with $x = 0$ representing 2003. Use this equation to approximate the number of instruments sold in each year from 2003 through 2006.

**27.** 2003 _____

2004 _____

2005 _____

2006 _____

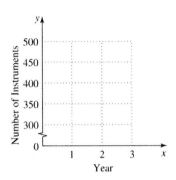

**28.** Suppose that the demand and price for a certain model of calculator are related by the equation $y = 45 - \frac{3}{5}x$, where $y$ is the price (in dollars) and $x$ is the demand (in thousands of calculators). Assuming that this model is valid for a demand up to 50,000 calculators, find the price at each of the following levels of demand.

(a)     0 calculators
(b)     5000 calculators
(c)     20,000 calculators
(d)     45,000 calculators

**28.** (a) _____

(b) _____

(c) _____

(d) _____

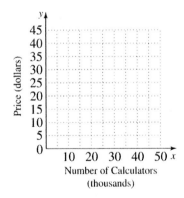

**351**

29. Every year sea turtles return to a certain group of islands to lay eggs. The number of turtle eggs that hatch can be approximated by the equation $y = -70x + 3260$, where $y$ is the number of eggs that hatch and $x = 0$ representing 1990. Use this equation to find the number of eggs that hatched in 1995, 2000, and 2005. Estimate the number of eggs that will hatch in 2015.

29. 1995 _____

    2000 _____

    2005 _____

    2015 _____

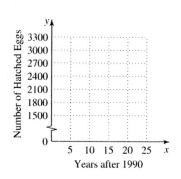

30. According to *The Old Farmer's Almanac*, the temperature in degrees Celsius can be determined by the equation $y = \dfrac{1}{3}x + 4$, where $x$ is the number of cricket chirps in 25 seconds and $y$ is the temperature in degrees Celsius. Use this equation to find the temperature when there are 48 chirps, 54 chirps, 60 chirps, and 66 chirps.

30. 48 _____

    54 _____

    60 _____

    66 _____

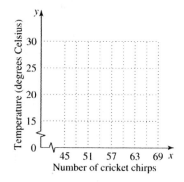

# Chapter 11 GRAPHS OF LINEAR EQUATIONS AND INEQUALITIES IN TWO VARIABLES

### 11.3   Slope of a Line

| Learning Objectives |
| --- |
| 1   Find the slope of a line given two points. |
| 2   Find the slope from the equation of a line. |
| 3   Use slope to determine whether two lines are parallel, perpendicular, or neither. |

### Key Terms

Use the vocabulary terms listed below to complete each statement in exercises 1–5.

> **rise**          **run**          **slope**          **parallel lines**
>
> **perpendicular lines**

1.   Two lines that intersect in a 90° angle are called _____.

2.   The _____ of a line is the ratio of the change in $y$ compared to the change in $x$ when moving along the line from one point to another.

3.   The vertical change between two different points on a line is called the
     _____.

4.   Two lines in a plane that never intersect are called _____.

5.   The horizontal change between two different points on a line is called the
     _____.

### Objective 1   Find the slope of a line given two points.

*Find the slope of the line through the given points.*

1.   (4, 3) and (3, 5)                              1. _____

2.   (5, −2) and (2, 7)                             2. _____

3.   (7, 2) and (−7, 3)                             3. _____

**4.**    $(-6, -6)$ and $(2, -6)$                              **4.** _____

**5.**    $(0, 0)$ and $(3, 5)$                                 **5.** _____

**6.**    $(-4, 6)$ and $(-4, -1)$                              **6.** _____

**7.**    $(-4, -7)$ and $(-2, 1)$                              **7.** _____

**8.**    $(-3, 3)$ and $(6, 3)$                                **8.** _____

**9.**    $(5, -2)$ and $(-5, 2)$                               **9.** _____

**10.**    $(4, 9)$ and $(-4, 1)$                               **10.** _____

**Objective 2    Find the slope from the equation of a line.**

*Find the slope of each line.*

**11.**    $y = \frac{1}{2}x + 5$                               **11.** _____

**12.**    $y = -5x$                                            **12.** _____

**13.**    $7y - 4x = 11$                                    **13.** _____

**14.**    $4x - 3y = 0$                                     **14.** _____

**15.**    $y = -4$                                          **15.** _____

**16.**    $3y = 2x - 1$                                     **16.** _____

**17.**    $y = -\frac{2}{5}x - 4$                           **17.** _____

**18.**    $x = 0$                                           **18.** _____

**19.**    $2x + 7y = 7$                                     **19.** _____

**20.**    $7y - 4x = 11$                                    **20.** _____

**Objective 3**    **Use slope to determine whether two lines are parallel, perpendicular, or neither.**

*In each pair of equations, give the slope of each line, and then determine whether the two lines are* **parallel**, **perpendicular**, *or* **neither**.

**21.**    $y = -5x - 2$

        $y = 5x + 11$

**21.** _____

**22.**    $y = 4x + 4$

        $y = 3 - \frac{1}{4}x$

**22.** _____

**23.**    $-x + y = -7$

        $x - y = -3$

**23.** _____

**24.**    $2x + 2y = 7$

        $2x - 2y = 5$

**24.** _____

**25.**    $4x + 2y = 8$

        $x + 4y = -3$

**25.** _____

**26.**     $9x + 3y = 2$

           $x - 3y = 5$

**26.** _____

**27.**     $4x + 2y = 7$

           $5x + 3y = 11$

**27.** _____

**28.**     $y = 9$

           $x = 0$

**28.** _____

**29.**     $8x + 2y = 7$

           $x = 3 - y$

**29.** _____

**30.**     $y = 6$

           $y + 2 = 9$

**30.** _____

# Chapter 11 GRAPHS OF LINEAR EQUATIONS AND INEQUALITIES IN TWO VARIABLES

## 11.4   Equations of Lines

| Learning Objectives |
| --- |
| 1      Write an equation of a line given its slope and *y*-intercept. |
| 2      Graph a line given its slope and a point on the line. |
| 3      Write an equation of a line given its slope and any point on the line. |
| 4      Write an equation of a line given two points on the line. |
| 5      Find an equation of a line that fits a data set. |

## Key Terms

Use the vocabulary terms listed below to complete each statement in exercises 1–3.

**slope-intercept form          point-slope form        standard form**

1.    A linear equation in the form $y - y_1 = m(x - x_1)$ is written in

_____.

2.    A linear equation in the form $Ax + By = C$ is written in

_____.

3.    A linear equation in the form $y = mx + b$ is written in

_____.

## Objective 1   Write an equation of a line given its slope and *y*-intercept.

*Write the slope-intercept form equation of the line with the given slope and y-intercept.*

1.    $m = \frac{3}{2};\ b = -\frac{2}{3}$

1. _____

2.    $m = -7;\ b = -2$

2. _____

3.    slope $-4$; *y*-intercept $(0,0)$

3. _____

4.    slope $0$; *y*-intercept $(0,-4)$

4. _____

*Use the geometric interpretation of slope to find the slope of each line. Then, by identifying the y-intercept from the graph, write the slope-intercept form of the equation of the line.*

**5.**

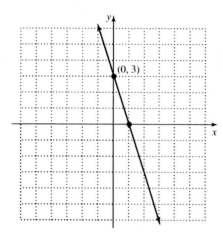

**5.** _____

**6.**

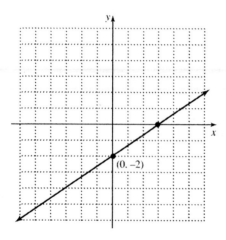

**6.** _____

## Objective 2   Graph a line given its slope and a point on the line.

*Graph the line passing through the given point and having the given slope.*

**7.**   $(4, -2)$; $m = -1$

**7.**

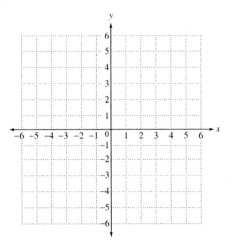

**8.**   $(-3, -2)$; $m = \frac{2}{3}$

**8.**

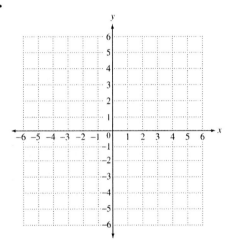

**9.**   $(-2, -2)$; $m = 0$

**9.**

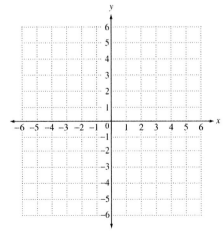

**10.**    $(1, -3); \; m = -\frac{5}{2}$

**10.**

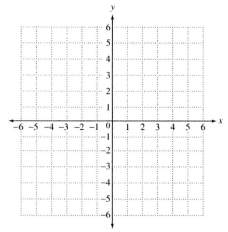

**11.**    $(2, 2); \; m = \frac{1}{3}$

**11.**

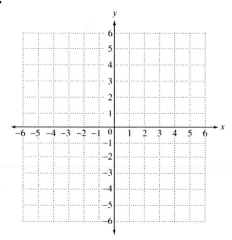

**12.**    $(-3, -1); \;$ undefined slope

**12.**

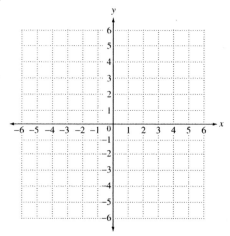

**Objective 3     Write an equation of a line given its slope and any point on the line.**

*Write an equation for the line passing through the given point and having the given slope. Write the equations in slope-intercept form, if possible.*

13.     $(-4,-2)$; undefined slope                    13. _____

14.     $(-3,4)$; $m = -\frac{3}{5}$                    14. _____

15.     $(-4,-7)$; $m = \frac{4}{3}$                    15. _____

16.     $(0,0)$; $m = 0$                    16. _____

17.     $(-1,2)$; $m = \frac{2}{3}$                    17. _____

18.     $(-4,-3)$; $m = -2$                    18. _____

19.     $(2,2)$; $m = -\frac{3}{2}$                    19. _____

## Objective 4    Write an equation of a line given two points on the line.

*Write an equation for the line passing through each pair of points. Write the equations in standard form.*

**20.**    $(-2,1)$ and $(3,11)$

**20.** _____

**21.**    $(2,3)$ and $(-2,-3)$

**21.** _____

**22.**    $\left(-\frac{4}{5},\frac{1}{8}\right)$ and $\left(-\frac{8}{5},-\frac{3}{8}\right)$

**22.** _____

**23.**    $\left(\frac{1}{2},\frac{2}{3}\right)$ and $\left(-\frac{3}{2},2\right)$

**23.** _____

**24.**    $(2,6)$ and $(-4,6)$

**24.** _____

**25.**    $(3,-4)$ and $(2,7)$.

**25.** _____

**26.**    $(-2,-4)$ and $(-2,7)$.

**26.** _____

## Objective 5    Find an equation of a line that fits a data set.

*Solve each problem.*

**27.**    The table shows the average annual telephone
expenditures for residential and pay telephones from
2001 to 2006, where year 0 represents 2001.
(Source: http://www.bls.gov/cex/cellphones.htm)

| Year | Annual Telephone Expenditures |
|------|-------------------------------|
| 0 | $686 |
| 2 | $620 |
| 3 | $592 |
| 4 | $570 |
| 5 | $542 |

(a)    Write five ordered pairs for the data.

(b)    Plot the ordered pairs.

(c)    Using the first and last points, find the equation
of a line that approximates the data. Write the
equation in slope-intercept form.

(d)    Use the equation from part (c) to predict the
annual telephone expenditures in 2010.

**27.** (a) _____

_____

(b)

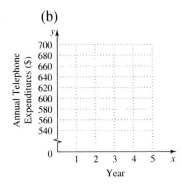

(c) _____

(d) _____

**28.** The table shows the number of internet users in the world from 1998 to 2005, where year 0 represents 1998. (Source: www.internetworldstats.com)

| Year | Number of Internet Users (millions) |
|------|-------------------------------------|
| 0    | 147                                 |
| 2    | 361                                 |
| 4    | 587                                 |
| 6    | 817                                 |
| 8    | 1093                                |

(a) Write five ordered pairs for the data.

(b) Plot the ordered pairs.

(c) Using the first and last points, find the equation of a line that approximates the data. Write the equation in slope-intercept form.

(d) Use the equation from part (c) to predict the number of internet users in 2010.

**28.** (a) _____

_____

(b)

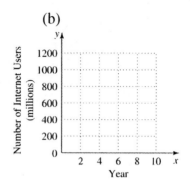

(c) _____

(d) _____

**29.** The table shows the U.S. municipal solid waste recycling percents since 1985, where year 0 represents 1985. (Source: www.epa.gov/epaoswer non-hw/muncpl/pubs/msw06.pdf)

| Year | Recycling Percent |
|------|-------------------|
| 0 | 10.1 |
| 5 | 16.2 |
| 10 | 26.0 |
| 15 | 29.1 |
| 20 | 32.5 |

(a) Write five ordered pairs for the data.

(b) Plot the ordered pairs.

(c) Using the first and last points, find the equation of a line that approximates the data. Write the equation in slope-intercept form.

(d) Use the equation from part (c) to predict the percent of municipal solid waste recycling in the year 2015.

**29.** (a) _____

_____

(b)

(c) _____

(d) _____

Name:                                    Date:
Instructor:                              Section:

**30.**   The table shows the approximate consumer
         expenditures for food in the U.S. in billions of
         dollars for selected years, where year 0 represents
         1985. (Source: http://www.ers.usda.gov/briefing/
         CPIFoodAndExpenditures/Data/table2.htm)

| Year | Food Expenditures (billions of dollars) |
|------|------------------------------------------|
| 0    | 233 |
| 5    | 298 |
| 10   | 343 |
| 15   | 417 |
| 20   | 515 |

(a)  Write five ordered pairs for the data.

(b)  Plot the ordered pairs.

(c)  Using the first and last points, find the equation
     of a line that approximates the data. Write the
     equation in slope-intercept form.

(d)  Use the equation from part (c) to predict the
     approximate food expenditures in the year 2012.

**30.** (a) _____

         _____

(b)

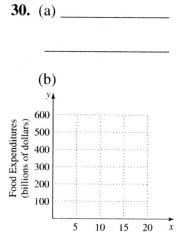

(c) _____

(d) _____

# Chapter 11 GRAPHS OF LINEAR EQUATIONS AND INEQUALITIES IN TWO VARIABLES

## 11.5    Graphing Linear Inequalities in Two Variables

| **Learning Objectives** |
| --- |
| 1    Graph linear inequalities. |
| 2    Graph an inequality with boundary through the origin. |

### Key Terms

Use the vocabulary terms listed below to complete each statement in exercises 1–2.

**linear inequality in two variables**            **boundary line**

1.    In the graph of a linear inequality, the _____ separates the region that satisfies the inequality from the region that does not satisfy the inequality.

2.    An inequality that can be written in the form $Ax + By < C$, $Ax + By > C$, $Ax + By \leq C$, or $Ax + By \geq C$ is called a_____.

### Objective 1    Graph linear inequalities.

*Graph each linear inequality.*

1.    $y \geq x - 1$

1.

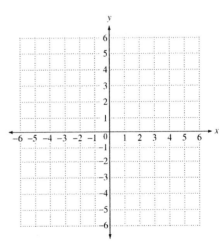

**2.**     $y > -x + 2$

**2.**

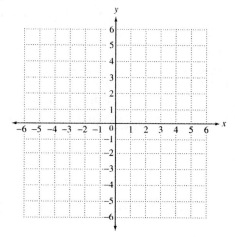

**3.**     $3x - 2y \leq 6$

**3.**

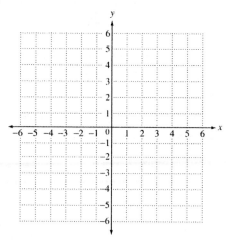

**4.**     $5x + 4y > 20$

**4.**

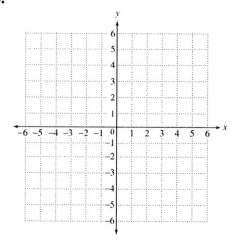

**5.**   $x - 4 \leq -1$                   **5.**

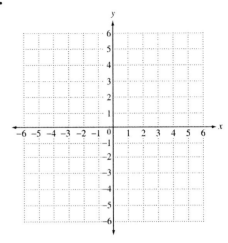

**6.**   $2x + 5y > -10$                   **6.**

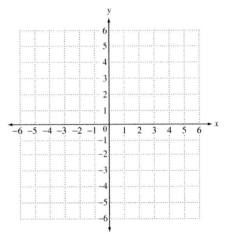

**7.**   $2x + 5y \leq -8$                 **7.**

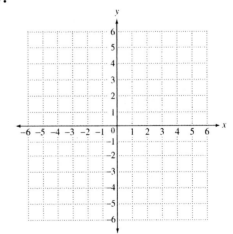

**8.**  $y \leq -\frac{1}{2}x + 6$

**8.**

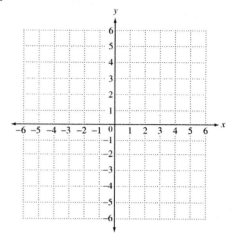

**9.**  $5x - 2y + 10 < 0$

**9.**

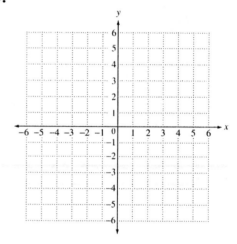

**10.**  $y \leq -\frac{2}{5}x + 2$

**10.**

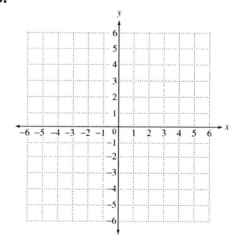

**11.**   $3x - 5y > -15$                          **11.**

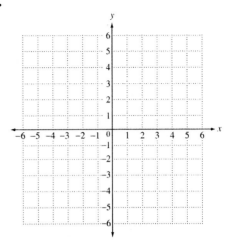

**12.**   $y \geq -1$                                **12.**

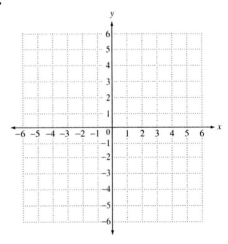

**13.**   $3x + 2y \leq -6$                          **13.**

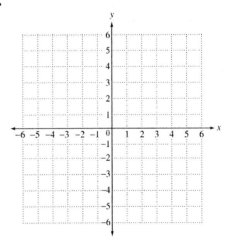

**14.**    $2 - 3y > x$

**14.**

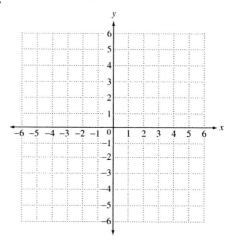

**15.**    $3x - 4y - 12 > 0$

**15.**

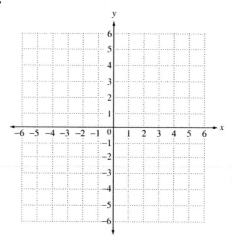

## Objective 2    Graph an inequality with boundary through the origin.

*Graph each linear inequality.*

**16.**    $y \geq 3x$

**16.**

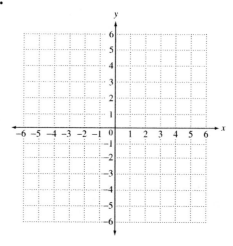

**17.** $y \leq \frac{2}{5}x$

**17.**

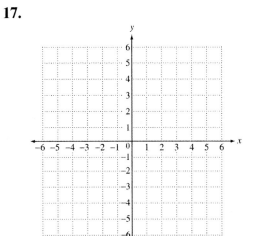

**18.** $y \geq \frac{1}{3}x$

**18.**

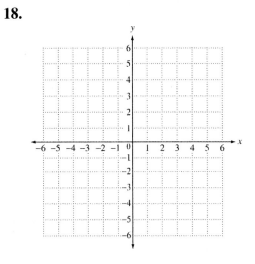

**19.** $y \geq x$

**19.**

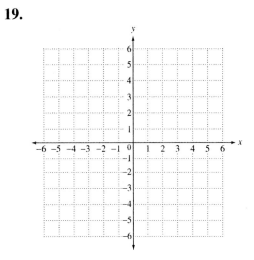

**20.** $3x - 4y \geq 0$

**20.**

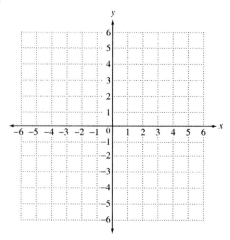

**21.** $x \geq -4y$

**21.**

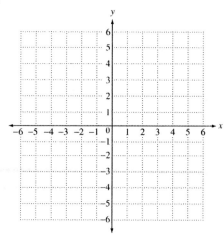

**22.** $x < 2y$

**22.**

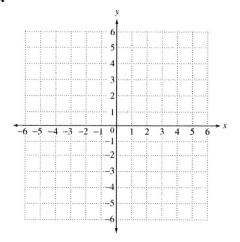

**23.**    $x < -2y$

**23.**

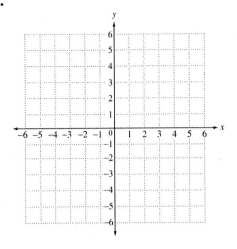

**24.**    $x > 4y$

**24.**

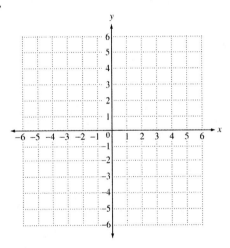

**25.**    $3x - 2y < 0$

**25.**

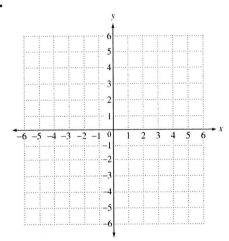

**26.**   $y \geq -\dfrac{1}{2}x$

**26.**

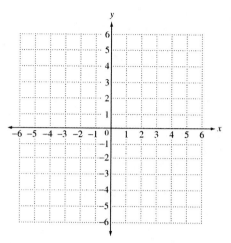

**27.**   $x < \dfrac{1}{3}y$

**27.**

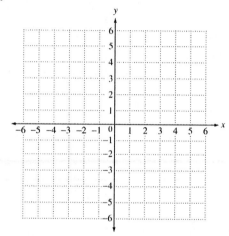

**28.**   $3x \geq 5y$

**28.**

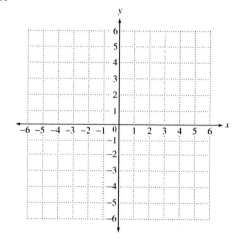

**29.**     $2y - 3x \geq 0$                    **29.**

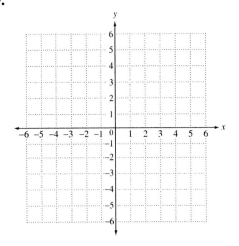

**30.**     $x > -\dfrac{1}{3}y$                    **30.**

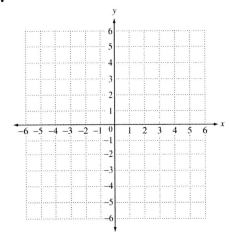

# Chapter 12 EXPONENTS AND POLYNOMIALS

### 12.1  Adding and Subtracting Polynomials

| **Learning Objectives** |
| --- |
| 1     Review combining like terms. |
| 2     Know the vocabulary for polynomials. |
| 3     Evaluate polynomials. |
| 4     Add polynomials. |
| 5     Subtract polynomials. |
| 6     Add and subtract polynomials with more than one variable. |

**Key Terms**

Use the vocabulary terms listed below to complete each statement in exercises 1–9.

| **term** | **like terms** | **polynomial** |
| --- | --- | --- |
| **descending powers** | | **degree of a term** |
| **degree of a polynomial** | | **monomial** |
| **binomial** | | **trinomial** |

1. The _____ is the sum of the exponents on the variables in that term.

2. A polynomial in $x$ is written in _____ if the exponents on $x$ decrease from left to right.

3. A _____ is a number, a variable, or a product or quotient of a number and one or more variables raised to powers.

4. A polynomial with exactly three terms is called a _____.

5. A _____ is a term, or the sum of terms, with whole number exponents.

6. A polynomial with exactly one term is called a _____.

7. The _____ is the highest degree of any term of the polynomial.

8. A _____ is a polynomial with exactly two terms.

9. Terms with exactly the same variables (including the same exponents) are called _____.

## Objective 1   Review combining like terms.

*In each polynomial, combine like terms whenever possible. Write the result with descending powers.*

**1.**   $7z^3 - 4z^3 + 5z^3 - 11z^3$

1. _____

**2.**   $-1.3z^7 + 0.4z^7 + 2.6z^8$

2. _____

**3.**   $-\frac{1}{2}r^3 + \frac{1}{3}r + \frac{1}{4}r^3 - \frac{1}{3}r$

3. _____

**4.**   $12x^3 + 7x^2 - 6x^3 + 5x^2$

4. _____

**5.**   $6c^3 - 9c^2 - 2c^2 + 14 + 3c^2 - 6c - 8 + 2c^3$

5. _____

## Objective 2   Know the vocabulary for polynomials.

*For each polynomial, first simplify, if possible, and write the resulting polynomial in descending powers of the variable. Then give the degree of this polynomial, and tell whether it is a* monomial, *a* binomial, *a* trinomial, *or* none of these.

**6.**   $3n^8 - n^2 - 2n^8$

6. _____

degree: _____

type: _____

**7.**   $\frac{7}{8}x^2 - \frac{3}{4}x - \frac{3}{8}x^2 + \frac{1}{4}x$

7. _____

degree: _____

type: _____

**8.** $10y^4 + 6y^2 - 12y^3 - 5y^4$

**8.** _____

degree: _____

type: _____

**9.** $-d^2 + 3.2d^3 - 5.7d^8 - 1.1d^5$

**9.** _____

degree: _____

type: _____

**10.** $-6c^4 - 6c^2 + 9c^4 - 4c^2 + 5c^5$

**10.** _____

degree: _____

type: _____

## Objective 3    Evaluate polynomials.

*Find the value of each polynomial* (a) *when x = –2 and* (b) *when x = 3.*

**11.** $2x^4 - 47$

**11. a.** _____

**b.** _____

**12.** $3x^3 + 4x - 19$

**12. a.** _____

**b.** _____

**13.** $-4x^3 + 10x^2 - 1$

**13. a.** _____

**b.** _____

**14.**   $x^4 - 3x^2 - 8x + 9$                    **14. a.** _____

                                                        **b.** _____

**15.**   $2x^4 + 7x^3 + x - 2$                    **15. a.** _____

                                                        **b.** _____

**Objective 4    Add polynomials.**

*Add.*

**16.**    $5m^4 + 2m^3 - 4$                       **16.** _____
          $\underline{-3m^4 + 5m^3 - 3}$

**17.**    $9m^3 + 4m^2 - 2m + 3$                  **17.** _____
          $\underline{-4m^3 - 6m^2 - 2m + 1}$

**18.**   $\left(x^2 + 6x - 8\right) + \left(3x^2 - 10\right)$        **18.** _____

**19.**   $\left(3r^3 + 5r^2 - 6\right) + \left(2r^2 - 5r + 4\right)$        **19.** _____

**20.**  $\left(3x^2 + 2x^4 - 3\right) + \left(8x^3 - 5x^4 - 6x^2\right)$

**20.** _____

## Objective 5    Subtract polynomials.

*Subtract.*

**21.**  $\left(-8w^3 + 11w^2 - 12\right) - \left(-10w^2 + 3\right)$

**21.** _____

**22.**  $\left(5a^4 - 6a^2 + 9a\right) - \left(a^3 - 19a - 1\right)$

**22.** _____

**23.**  $\left(8b^4 - 4b^3 + 7\right) - \left(2b^2 + b + 9\right)$

**23.** _____

**24.**  $\left(9x^3 + 7x^2 - 6x + 3\right) - \left(6x^3 - 6x + 1\right)$

**24.** _____

**25.**  $\left(7d^4 + 7d^2 - 8d + 12\right) - \left(10d^2 + 11d\right)$

**25.** _____

## Objective 6    Add and subtract polynomials with more than one variable.

*Add or subtract as indicated.*

**26.**    $\left(-2a^6 + 8a^4b - b^2\right) - \left(a^6 + 7a^4b + 2b^2\right)$          **26.** _____

**27.**    $\left(-4m^2n + 3n - 6m\right) - \left(2m + 7n + 4nm^2\right)$          **27.** _____

**28.**    $\left(4ab + 2bc - 9ac\right) + \left(3ca - 2cb - 9ba\right)$          **28.** _____

**29.**    $\left(2x^2y + 2xy - 4xy^2\right) + \left(6xy + 9xy^2\right) - \left(9x^2y + 5xy\right)$          **29.** _____

**30.**    $\left(.01ab + .03a^2 - .05b^2\right) - \left(-.08a^2 + .02b^2 + .01ab\right)$          **30.** _____

# Chapter 12 EXPONENTS AND POLYNOMIALS

### 12.2 The Product Rule and Power Rules for Exponents

| **Learning Objectives** |
| --- |
| 1    Use exponents. |
| 2    Use the product rule for exponents. |
| 3    Use the rule $\left(a^m\right)^n = a^{mn}$. |
| 4    Use the rule $(ab)^m = a^m b^m$. |
| 5    Use the rule $\left(\dfrac{a}{b}\right)^m = \dfrac{a^m}{b^m}$. |
| 6    Use combinations of the rules for exponents. |
| 7    Use the rules for exponents in a geometry application. |

### Key Terms

Use the vocabulary terms listed below to complete each statement in exercises 1–3.

       **exponential expression**      **base**        **power**

1.   $2^5$ is read "2 to the fifth _____".

2.   A number written with an exponent is called a(n)
     _____.

3.   The _____ is the number being multiplied repeatedly.

### Objective 1    Use exponents.

*Write each expression in exponential form and evaluate, if possible.*

1.   $\left(\frac{1}{3}\right)\left(\frac{1}{3}\right)\left(\frac{1}{3}\right)\left(\frac{1}{3}\right)\left(\frac{1}{3}\right)$                      1.   _____

2.   $(.5st)(.5st)(.5st)(.5st)$                          2.   _____

*Evaluate each exponential expression. Name the base and the exponent.*

3.   $(-4)^4$                                3.   _____

                                            base_____

                                            exponent _____

**4.** $-3^8$

**4.** _____

base _____

exponent _____

## Objective 2  Use the product rule for exponents.

*Use the product rule to simplify each expression, if possible. Write each answer in exponential form.*

**5.** $7^4 \cdot 7^3$

**5.** _____

**6.** $\left(\frac{1}{2}\right)^9 \cdot \left(\frac{1}{2}\right)$

**6.** _____

**7.** $\left(-2c^7\right)\left(-4c^8\right)$

**7.** _____

**8.** $\left(3k^7\right)\left(-8k^2\right)\left(-2k^9\right)$

**8.** _____

## Objective 3  Use the rule $\left(a^m\right)^n = a^{mn}$.

*Simplify each expression. Write all answers in exponential form.*

**9.** $\left(7^3\right)^4$

**9.** _____

**10.** $-\left(v^4\right)^9$

**10.** _____

**11.** $\left[\left(\frac{1}{3}\right)^3\right]^5$

**11.** _____

**12.** $\left[ (-3)^3 \right]^7$

12. _____

**Objective 4    Use the rule** $(ab)^m = a^m b^m$.

*Simplify each expression.*

**13.** $\left( \frac{1}{3} x^4 \right)^2$

13. _____

**14.** $\left( 5r^3 t^2 \right)^4$

14. _____

**15.** $\left( -.2a^4 b \right)^3$

15. _____

**16.** $\left( -2w^3 z^7 \right)^4$

16. _____

**Objective 5    Use the rule** $\left( \dfrac{a}{b} \right)^m = \dfrac{a^m}{b^m}$.

*Simplify each expression.*

**17.** $\left( -\dfrac{2x}{5} \right)^3$

17. _____

**18.** $\left( \dfrac{xy}{z^2} \right)^4$

18. _____

**19.** $\left( \dfrac{-2a}{b^2} \right)^7$

19. _____

**20.** $-\left(\dfrac{2a^3c}{5b^2}\right)^5$                    20. _____

## Objective 6    Use combinations of the rules for exponents.

*Simplify. Write all answers in exponential form.*

**21.** $\left(-x^3\right)^2\left(-x^5\right)^4$                    21. _____

**22.** $\left(2ab^2c\right)^5(ab)^4$                    22. _____

**23.** $\left(5x^2y^3\right)^7\left(5xy^4\right)^4$                    23. _____

**24.** $-\left(\dfrac{2a^3c}{5b^2}\right)^5$ $(b \neq 0)$                    24. _____

**25.** $\left(7a^2b^2c\right)^3\left(ab^3c^4\right)^4$                    25. _____

## Objective 7    Use the rules for exponents in a geometry application.

*Find a polynomial that represents the area of each figure.*

**26.**                                    26. _____

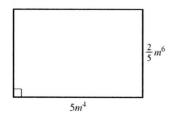

$\frac{2}{5}m^6$

$5m^4$

**27.**

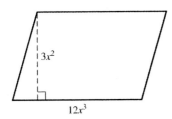

**27.** _____

**28.**

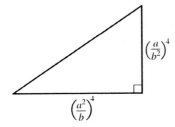

**28.** _____

**29.**

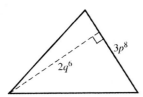

**29.** _____

**30.**

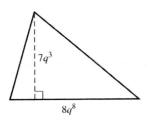

**30.** _____

Name: _____                    Date: _____

Instructor: _____              Section: _____

# Chapter 12 EXPONENTS AND POLYNOMIALS

### 12.3 Multiplying Polynomials

| Learning Objectives |
| --- |
| 1     Multiply a monomial and a polynomial. |
| 2     Multiply two polynomials. |
| 3     Multiply binomials by the FOIL method. |

### Key Terms

Use the vocabulary terms listed below to complete each statement in exercises 1–3.

**FOIL**          **outer product**          **inner product**

1. The _____ of $(2y-5)(y+8)$ is $-5y$.

2. _____ is a shortcut method for finding the product of two binomials.

3. The _____ of $(2y-5)(y+8)$ is $16y$.

### Objective 1  Multiply a monomial and a polynomial.

*Find each product.*

1. $\left(-2y^3\right)\left(-8y^4\right)$                                          1. _____

2. $7z\left(5z^3+2\right)$                                                  2. _____

3. $2m\left(3+7m^2+3m^3\right)$                                        3. _____

4. $4k^2\left(3+2k^3+6k^4\right)$                                        4. _____

5. $-6z\left(z^5+3z^3+4z+2\right)$                                     5. _____

**6.** $\quad -3y^2\left(2y^3 + 3y^2 - 4y + 11\right)$            **6.** _____

**7.** $\quad 7b^2\left(-5b^2 + 1 - 4b\right)$            **7.** _____

**8.** $\quad -4r^4\left(2r^2 - 3r + 2\right)$            **8.** _____

**9.** $\quad 8mn\left(4m^2 + 2mn + 7n^2\right)$            **9.** _____

**10.** $\quad -3r^2s^3\left(8r^2s^2 - 4rs + 2rs^2\right)$            **10.** _____

**Objective 2    Multiply two polynomials.**

*Find each product.*

**11.** $\quad (2x + 5)(3x + 4)$            **11.** _____

**12.** $\quad (3m - 5)(2m + 4)$            **12.** _____

**13.** $\quad (x + 3)\left(x^2 - 3x + 9\right)$            **13.** _____

**14.** $\quad (y + 4)\left(y^2 - 4y + 16\right)$            **14.** _____

**15.** $(r+3)\left(2r^2 - 3r + 5\right)$                              **15.** _____

**16.** $(3y-4)\left(3y^3 - 2y^2 - y + 4\right)$                   **16.** _____

**17.** $\left(2m^2 + 1\right)\left(3m^3 + 2m^2 - 4m\right)$      **17.** _____

**18.** $\left(3x^2 + x\right)\left(2x^2 + 3x - 4\right)$         **18.** _____

**19.** $\left(y^2 - 2y - 3\right)\left(y^2 - 2y - 3\right)$      **19.** _____

**20.** $\left(3x^3 + 3x^2 - 2x + 1\right)\left(2x^2 - x + 2\right)$   **20.** _____

**Objective 3    Multiply binomials by the FOIL method.**

*Find each product.*

**21.** $(3x + 2y)(2x - 3y)$                              **21.** _____

**22.** $(5a - b)(4a + 3b)$

**22.** _____

**23.** $(3 + 4a)(1 + 2a)$

**23.** _____

**24.** $(4x - 3y)(x + 2y)$

**24.** _____

**25.** $(2m + 3n)(-3m + 4n)$

**25.** _____

**26.** $\left(2v^2 + w^2\right)\left(v^2 - 3w^2\right)$

**26.** _____

**27.** $(x - .5)(x + .3)$

**27.** _____

**28.** $(2y + .1)(2y - .5)$

**28.** _____

**29.** $\left(x - \frac{1}{3}\right)\left(x - \frac{4}{3}\right)$

**29.** _____

**30.** $\left(z + \frac{4}{5}\right)\left(z - \frac{2}{5}\right)$

**30.** _____

# Chapter 12 EXPONENTS AND POLYNOMIALS

## 12.4   Special Products

| Learning Objectives |
|---|
| **1** Square binomials. |
| **2** Find the product of the sum and difference of two terms. |
| **3** Find greater powers of binomials. |

## Key Terms

Use the vocabulary terms listed below to complete each statement in exercises 1–2.

**conjugate**              **binomial**

1. A polynomial with two terms is called a _____.

2. The _____ of $a + b$ is $a - b$.

## Objective 1   Square binomials.

*Find each square by using the pattern for the square of a binomial.*

1. $(5y - 3)^2$                                      1. _____

2. $(2m + 5)^2$                                      2. _____

3. $(7 + x)^2$                                       3. _____

4. $(5 - 3y)^2$                                      4. _____

5. $(2p + 3q)^2$                                     5. _____

6. $(2m - 3p)^2$                                     6. _____

**7.** $(4y - .7)^2$

**8.** $\left(4x - \frac{1}{4}y\right)^2$

**9.** $\left(3x + \frac{1}{3}y\right)^2$

**10.** $\left(3a + \frac{1}{2}b\right)^2$

**Objective 2   Find the product of the sum and difference of two terms.**

*Find each product by using the pattern for the sum and difference of two terms.*

**11.** $(12 + x)(12 - x)$

**12.** $(8k + 5p)(8k - 5p)$

**13.** $(7x - 3y)(7x + 3y)$

**14.** $(2 + 3x)(2 - 3x)$

**15.** $(x + .2)(x - .2)$

**16.** $(9 - .4y)(9 + .4y)$

**17.** $\left(7m - \frac{3}{4}\right)\left(7m + \frac{3}{4}\right)$          **17.** _____

**18.** $\left(\frac{3}{4}s + \frac{7}{5}t\right)\left(\frac{3}{4}s - \frac{7}{5}t\right)$          **18.** _____

**19.** $\left(4x + \frac{7}{4}\right)\left(4x - \frac{7}{4}\right)$          **19.** _____

**20.** $\left(\frac{4}{7}t + 2u\right)\left(\frac{4}{7}t - 2u\right)$          **20.** _____

**Objective 3    Find greater powers of binomials.**

*Find each product.*

**21.** $(a - 3)^3$          **21.** _____

**22.** $(2x + 4)^3$          **22.** _____

**23.** $(4x + y)^3$          **23.** _____

**24.** $(.3x - .2y)^3$          **24.** _____

**25.** $\left(\frac{1}{2}t + 2u\right)^3$

**25.** _____

**26.** $(j+3)^4$

**26.** _____

**27.** $(x+2y)^4$

**27.** _____

**28.** $(3b-2)^4$

**28.** _____

**29.** $(2x-y)^4$

**29.** _____

**30.** $(4s+3t)^4$

**30.** _____

# Chapter 12 EXPONENTS AND POLYNOMIALS

## 12.5    Integer Exponents and the Quotient Rule

| **Learning Objectives** |
| --- |
| 1    Use 0 as an exponent. |
| 2    Use negative numbers as exponents. |
| 3    Use the quotient rule for exponents. |
| 4    Use combinations of rules. |

## Key Terms

Use the vocabulary terms listed below to complete each statement in exercises 1–3.

**exponent**                    **base**                    **product rule for exponents**

**power rule for exponents**

1.    The statement "If $m$ and $n$ are any integers, then $\left(a^m\right)^n = a^{mn}$ " is an example of the

_____.

2.    In the expression $a^m$, $a$ is the _____ and $m$ is the _____.

3.    The statement "If $m$ and $n$ are any integers, then $a^m \cdot a^n = a^{m+n}$ is an example of

the _____.

## Objective 1    Use 0 as an exponent.

*Evaluate each expression.*

1.    $-(-8)^0$                                                         1. _____

2.    $-12^0$                                                           2. _____

3.    $2^0 + 6^0$                                                       3. _____

4.    $\left(\frac{2}{3}\right)^0 + \left(\frac{1}{3}\right)^0 - 2^0$                4. _____

5.    $-15^0 - (-15)^0$                                                 5. _____

6.    $-r^0 \ (r \neq 0)$                                               6. _____

**7.** $\dfrac{0^8}{8^0}$

7. _____

## Objective 2    Use negative numbers as exponents.

*Evaluate or simplify each expression, and write it using only positive exponents. Assume that all variables represent nonzero real numbers.*

**8.**    $4^{-2}$

8. _____

**9.**    $\dfrac{2}{r^{-7}}$

9. _____

**10.**    $-2k^{-4}$

10. _____

**11.**    $-(-4)^{-4}$

11. _____

**12.**    $(m^2 n)^{-9}$

12. _____

**13.**    $3x^{-2} - \dfrac{6}{x^2}$

13. _____

**14.**    $\dfrac{2x^{-4}}{3y^{-7}}$

14. _____

**15.**    $\dfrac{2^{-4}}{8^{-2}}$

15. _____

## Objective 3    Use the quotient rule for exponents.

*Use the quotient rule to simplify each expression, and write it using only positive exponents. Assume that all variables represent nonzero real numbers.*

**16.**    $\dfrac{12^{-7}}{12^{-6}}$

16. _____

**17.** $\dfrac{2^4 \cdot x^2}{2^5 \cdot x^8}$

**17.** _____

**18.** $\dfrac{4k^7 m^{10}}{8k^3 m^5}$

**18.** _____

**19.** $\dfrac{12x^9 y^5}{12^4 x^3 y^7}$

**19.** _____

**20.** $\dfrac{a^4 b^3}{a^{-2} b^{-3}}$

**20.** _____

**21.** $\dfrac{3^{-1} m^{-4} p^6}{3^4 m^{-1} p^{-2}}$

**21.** _____

**22.** $\dfrac{8b^{-3} c^4}{8^{-4} b^{-7} c^{-3}}$

**22.** _____

## Objective 4  Use combinations of rules.

*Simplify each expression, and write it using only positive exponents. Assume that all variables represent nonzero real numbers.*

**23.** $\dfrac{(2y)^{-4}}{(3y)^{-2}}$

**23.** _____

**24.** $\left(2p^{-3} q^2\right)^2 \left(4p^4 q^{-1}\right)^{-1}$

**24.** _____

**25.**  $(9xy)^7 (9xy)^{-8}$

**25.** _____

**26.**  $\dfrac{c^{10}\left(c^2\right)^3}{\left(c^3\right)^3\left(c^2\right)^{-9}}$

**26.** _____

**27.**  $\dfrac{\left(a^{-1}b^{-2}\right)^{-4}\left(ab^2\right)^6}{\left(a^3b\right)^{-2}}$

**27.** _____

**28.**  $\left(\dfrac{k^3t^4}{k^2t^{-1}}\right)^{-4}$

**28.** _____

**29.**  $\dfrac{\left(3^{-2}x^{-5}y\right)^{-4}\left(2x^2y^{-4}\right)^2}{\left(2x^{-2}y^2\right)^{-2}}$

**29.** _____

**30.**  $\dfrac{\left(4a^{-1}b^4\right)^{-3}\left(4a^2b\right)^{-1}}{\left(2a^{-3}b^2\right)^{-2}}$

**30.** _____

# Chapter 12 EXPONENTS AND POLYNOMIALS

### 12.6   Dividing a Polynomial by a Monomial

| Learning Objectives |
| --- |
| 1     Divide a polynomial by a monomial. |

### Key Terms

Use the vocabulary terms listed below to complete each statement in exercises 1–3.

      **quotient**       **dividend**       **divisor**

1.   In the division $\dfrac{5x^5 - 10x^3}{5x^2} = x^3 - 2x$, the expression $5x^5 - 10x^3$ is the

      _____.

2.   In the division $\dfrac{5x^5 - 10x^3}{5x^2} = x^3 - 2x$, the expression $x^3 - 2x$ is the

      _____.

3.   In the division $\dfrac{5x^5 - 10x^3}{5x^2} = x^3 - 2x$, the expression $5x^2$ is the

      _____.

### Objective 1   Divide a polynomial by a monomial.

*Perform each division.*

1.   $\dfrac{16a^5 - 24a^3}{8a^2}$                                               1. _____

2.   $\left(20x^4 - 10x^2\right) \div \left(2x\right)$                       2. _____

3.   $\left(20a^3 - 9a\right) \div \left(4a\right)$                         3. _____

4.   $\dfrac{12x^6 + 28x^5 + 20x^3}{4x^2}$                          4. _____

**5.** $\dfrac{6p^4 + 18p^7}{6p^4}$

    **5.** _____

**6.** $\dfrac{16x^4 - 12x^3 + 8x^2}{4x^3}$

    **6.** _____

**7.** $\dfrac{6y^5 + 9y^8 - 21y^{10}}{3y^5}$

    **7.** _____

**8.** $\dfrac{24w^8 + 12w^6 - 18w^4}{-6w^5}$

    **8.** _____

**9.** $\dfrac{40p^4 - 35p^3 - 15p}{5p^2}$

    **9.** _____

**10.** $\dfrac{2.4a^2b^2 + .6ab^2 + 3a^2b}{.3ab}$

    **10.** _____

**11.** $\dfrac{9r^2s + 18rs^2 - 27s^3}{-27rs^2}$

    **11.** _____

**12.** $\dfrac{30x^3y^3 - 45x^2y^2 + 15xy}{-75xy^2}$

    **12.** _____

**13.** $\dfrac{32a^5 - 6a^4b + 24a^3b^2}{-8a^2}$

    **13.** _____

**14.** $\dfrac{3.2st^2 + .8s^2t + .4s^2t^2}{-.4st}$  14. _____

**15.** $\dfrac{-28p^5 - 21p^3 - 35p^2 + p}{7p^3}$  15. _____

**16.** $\dfrac{5x^2y^4 - 30x^4y^3 + 30x^5y^2}{-5x^2y^2}$  16. _____

**17.** $\left(6z^5 + 27z^3 - 12z + 10\right) \div (3z)$  17. _____

**18.** $\left(9x^4 + 24x^3 - 48x + 12\right) \div (3x)$  18. _____

**19.** $\left(m^2 + 7m - 42\right) \div (2m)$  19. _____

**20.** $\dfrac{70q^4 - 40q^2 + 10q}{10q^2}$  20. _____

**21.** $\dfrac{2y^9 + 8y^6 - 41y^3 - 12}{y^3}$  21. _____

**22.** $\dfrac{12z^5 + 28z^4 - 8z^3 + 3z}{4z^3}$  22. _____

**23.** $\dfrac{48x + 64x^4 + 2x^8}{4x}$

                                                        **23.** _____

**24.** $\dfrac{-25u^3v + 20u^2v^2 - 45uv^3}{5uv}$

                                                        **24.** _____

**25.** $\dfrac{21y^2 - 14y + 42}{-7y^2}$

                                                        **25.** _____

**26.** $\dfrac{39m^4 - 12m^3 + 15}{-3m^2}$

                                                        **26.** _____

**27.** $\dfrac{-20d^4 - 8d^3 + 14d^2 + 8}{-2d^2}$

                                                        **27.** _____

**28.** $\left(15x^5 - 10x^4 - 10x^2 + 4\right) \div \left(-5x\right)$

                                                        **28.** _____

**29.** $\dfrac{-12y^4 - 15y^3 + 2y}{-3y^2}$

                                                        **29.** _____

**30.** $\dfrac{18r^4 - 12r^3 + 36r^2 - 12}{6r}$

                                                        **30.** _____

# Chapter 12 EXPONENTS AND POLYNOMIALS

### 12.7  Dividing a Polynomial by a Polynomial

| **Learning Objectives** |
| --- |
| 1  Divide a polynomial by a polynomial. |
| 2  Apply division to a geometry problem. |

### Key Terms

Use the vocabulary terms listed below to complete each statement in exercises 1–3.

**quotient     dividend     divisor**

1.  In the division $\dfrac{6x^2 - 9x - 12}{2x - 5} = 3x + 3 + \dfrac{3}{2x - 5}$, the expression $2x - 5$ is the

    _____.

2.  In the division $\dfrac{6x^2 - 9x - 12}{2x - 5} = 3x + 3 + \dfrac{3}{2x - 5}$, the expression $3x + 3 + \dfrac{3}{2x - 5}$ is

    the _____.

3.  In the division $\dfrac{6x^2 - 9x - 12}{2x - 5} = 3x + 3 + \dfrac{3}{2x - 5}$, the expression $6x^2 - 9x - 12$ is

    the _____.

### Objective 1  Divide a polynomial by a polynomial.

*Perform each division.*

1.  $\dfrac{18a^2 - 9a - 5}{3a + 1}$

    1.  _____

2.  $\left(r^2 - r - 20\right) \div \left(r - 5\right)$

    2.  _____

**3.** $\dfrac{p^2 + 5p - 24}{p - 3}$

**3.** _____

**4.** $\left(9w^2 + 12w + 4\right) \div \left(3w + 2\right)$

**4.** _____

**5.** $\dfrac{81a^2 - 1}{9a + 1}$

**5.** _____

**6.** $\dfrac{4x^2 - 25}{2x - 5}$

**6.** _____

**7.** $\left(2a^2 - 11a + 16\right) \div \left(2a + 3\right)$

**7.** _____

**8.**     $\dfrac{5w^2 - 22w + 4}{w - 4}$

**8.** _____

**9.**     $\dfrac{9m^2 - 18m + 16}{3m - 4}$

**9.** _____

**10.**     $\dfrac{-6x^2 + 23x - 20}{2x - 5}$

**10.** _____

**11.**     $\dfrac{5b^2 + 32b + 3}{b + 7}$

**11.** _____

**12.** $\dfrac{12p^3 - 28p^2 + 21p - 5}{6p - 5}$

**12.** _____

**13.** $\dfrac{12y^3 - 11y^2 + 9y + 18}{4y + 3}$

**13.** _____

**14.** $\dfrac{2z^3 - 7z^2 + 3z + 2}{2z + 3}$

**14.** _____

**15.** $\dfrac{6m^3 + 7m^2 - 13m + 16}{3m + 2}$

**15.** _____

**16.**   $\left(27p^4 - 36p^3 - 6p^2 + 23p - 20\right) \div \left(3p - 4\right)$        **16.** _____

**17.**   $\left(3x^3 - 11x^2 + 25x - 25\right) \div \left(x^2 - 3x - 5\right)$        **17.** _____

**18.**   $\dfrac{6x^4 - 12x^3 + 13x^2 - 5x - 1}{2x^2 + 3}$        **18.** _____

**19.**   $\dfrac{12y^5 - 8y^4 - y^3 + 2y^2 - 5}{4y^2 - 3}$        **19.** _____

**20.**  $\dfrac{2a^4 + 5a^2 + 3}{2a^2 + 3}$

**20.** _____

**21.**  $\dfrac{y^3 + 1}{y + 1}$

**21.** _____

**22.**  $\dfrac{b^4 - 1}{b^2 - 1}$

**22.** _____

**23.**  $\dfrac{3x^4 + 2x^3 - 2x^2 - 2x - 1}{x^2 - 1}$

**23.** _____

**24.** $\dfrac{6x^5 + 7x^4 - 7x^3 + 7x + 2}{3x + 2}$

**24.** _____

**25.** $\dfrac{32x^5 - 243}{2x - 3}$

**25.** _____

**26.** $\dfrac{6y^6 + 8y^5 - 8y^4 - 3y^3 - 4y^2 + 4y + 3}{2y^3 - 1}$

**26.** _____

**Objective 2  Apply division to a geometry problem.**

*Work each problem.*

**27.**  The area of a rectangle is given by
$6r^3 - 5r^2 + 16r - 5$ square units, and the width is
$3r - 1$ units. What is the length of the rectangle?

**27.** _____

**28.** The area of a rectangle is given by
$12p^3 - 7p^2 + 5p - 1$ square units, and the width is
$4p - 1$ units. What is the length of the rectangle?

28.

**29.** The area of a parallelogram is given by
$4y^3 - 44y - 600$ square units, and the height is
$y - 6$ units. What is the base of the parallelogram?

29.

**30.** The area of a parallelogram is given by
$3t^3 + 16t^2 - 32t - 64$ square units, and the base is
$t^2 + 4t - 16$ units. What is the height of the
parallelogram?

30.

# Chapter 12 EXPONENTS AND POLYNOMIALS

### 12.8    An Application of Exponents: Scientific Notation

| Learning Objectives |
| --- |
| 1      Express numbers in scientific notation. |
| 2      Convert numbers in scientific notation to numbers without exponents. |
| 3      Use scientific notation in calculations. |

### Key Terms

Use the vocabulary terms listed below to complete each statement in exercises 1–3.

**scientific notation            quotient rule            power rule**

1.    A number written as $a \times 10^n$, where $1 \le |a| < 10$ and $n$ is an integer, is written in

    _____.

2.    The statement "If $m$ and $n$ are any integers and $b \ne 0$, then $\left(\dfrac{a}{b}\right)^m = \dfrac{a^m}{b^m}$" is an

    example of the _____.

3.    The statement "If $m$ and $n$ are any integers and $b \ne 0$, then $\dfrac{a^m}{a^n} = a^{m-n}$" is an

    example of the _____.

### Objective 1    Express numbers in scientific notation.

*Write each number in scientific notation.*

1.    325                                              1. _____

2.    4579                                             2. _____

3.    23,651                                           3. _____

4.    −38,600,000                                      4. _____

5.    9,540,000                                        5. _____

6.    −429,600,000,000                                 6. _____

7.    0.0503                                           7. _____

**8.**   0.007068                                    **8.** _____

**9.**   −0.0002208                                  **9.** _____

**10.**  0.00000476                                  **10.** _____

**Objective 2    Convert numbers in scientific notation to numbers without exponents.**

*Write each number in scientific notation.*

**11.**   $7.2 \times 10^7$                          **11.** _____

**12.**   $-2.45 \times 10^6$                        **12.** _____

**13.**   $2.3 \times 10^4$                          **13.** _____

**14.**   $4.5 \times 10^7$                          **14.** _____

**15.**   $6.4 \times 10^{-3}$                       **15.** _____

**16.**   $7.24 \times 10^{-4}$                      **16.** _____

**17.**   $4.007 \times 10^{-2}$                     **17.** _____

**18.**   $4.752 \times 10^{-1}$                     **18.** _____

**19.**   $-4.02 \times 10^0$                        **19.** _____

**20.**   $-9.11 \times 10^{-4}$                     **20.** _____

**Objective 3    Use scientific notation in calculations.**

*Perform the indicated operations, and write the answers in scientific notation.*

**21.**   $\left(2.3 \times 10^4\right) \times \left(1.1 \times 10^{-2}\right)$        **21.** _____

**22.**    $\dfrac{9.39 \times 10^1}{3 \times 10^3}$

22. _____

**23.**    $\left(6 \times 10^4\right) \times \left(3 \times 10^5\right) \div \left(9 \times 10^7\right)$

23. _____

**24.**    $\left(3 \times 10^4\right) \times \left(4 \times 10^2\right) \div \left(2 \times 10^3\right)$

24. _____

**25.**    $\dfrac{\left(7.5 \times 10^6\right) \times \left(4.2 \times 10^{-5}\right)}{\left(6 \times 10^4\right) \times \left(2.5 \times 10^{-3}\right)}$

25. _____

**26.**    $\dfrac{\left(2.1 \times 10^{-3}\right) \times \left(4.8 \times 10^4\right)}{\left(1.6 \times 10^6\right) \times \left(7 \times 10^{-6}\right)}$

26. _____

*Work each problem. Give answers without exponents.*

**27.**    There are about $6 \times 10^{23}$ atoms in a mole of atoms. How many atoms are there in $8.1 \times 10^{-5}$ mole?

27. _____

**28.** The earth has a mass of $6 \times 10^{24}$ kilograms and a volume of $1.1 \times 10^{21}$ cubic meters. What is the Earth's density in kilograms per cubic meter? Round to the nearest hundred.

**28.** _____

**29.** A light-year is the distance that light travels in one year. The speed of light is about $3 \times 10^5$ km per second. How many kilometers are in a light-year?

**29.** _____

**30.** The Sahara desert covers approximately $3.5 \times 10^6$ square miles. Its sand is, on average, 12 feet deep.

    **a.** Find the volume, in cubic feet, of sand in the Sahara. $\left(\text{Hint: } 1 \text{ mi}^2 = 5280^2 \text{ ft}^2\right)$ Round your answer to two decimal places.

**30. a.** _____

    **b.** _____

    **b.** The volume of a single grain of sand is approximately $1.3 \times 10^{-9}$ cubic feet. About how many grains of sand are in the Sahara?

# Chapter 13 FACTORING AND APPLICATIONS

### 13.1    Factors; The Greatest Common Factor

**Learning Objectives**
1    Find the greatest common factor of a list of numbers.
2    Find the greatest common factor of a list of variable terms.
3    Factor out the greatest common factor.
4    Factor by grouping.

**Key Terms**

Use the vocabulary terms listed below to complete each statement in exercises 1–4.

> **factor**          **factored form**          **greatest common factor (GCF)**
>
> **factoring**

1.   The process of writing a polynomial as a product is called _____.

2.   An expression is in _____ when it is written as a product.

3.   The _____ is the largest quantity that is a factor of each of a group of quantities.

4.   An expression $A$ is a _____ of an expression $B$ if $B$ can be divided by $A$ with 0 remainder.

### Objective 1    Find the greatest common factor of a list of numbers.

*Find the greatest common factor for each group of numbers.*

1.   60, 75, 120                                      1. _____

2.   108, 48, 84                                      2. _____

3.   9, 18, 24, 48                                    3. _____

4.   70, 126, 42, 56                                  4. _____

5.   84, 280, 112                                     5. _____

**6.**   56, 21, 49                                            **6.** _____

**7.**   42, 48, 72                                            **7.** _____

**Objective 2    Find the greatest common factor of a list of variable terms.**

*Find the greatest common factor for each list of terms.*

**8.**   $-15ab^2$, $-45a^3b^4$, $70ab^3$                    **8.** _____

**9.**   $12ab^3$, $18a^2b^4$, $26ab^2$, $32a^2b^2$         **9.** _____

**10.**  $6k^2m^4n^5$, $8k^3m^7n^4$, $k^4m^8,n^7$           **10.** _____

**11.**  $29w^3x^7y^4$, $w^4x^5y^7$, $58w^2x^9,y^5$         **11.** _____

**12.**  $45a^7y^4$, $75a^3y^2$, $-90a^2y$, $30a^4y^3$      **12.** _____

**13.**  $9xy^4,72x^4y^7$, $27xy^2$, $108x^2y^5$            **13.** _____

**14.**  $-72u^2v^3$, $-54uv^2$, $-63uv^4$                  **14.** _____

## Objective 3    Factor out the greatest common factor.

*Factor out the greatest common factor or a negative common factor if the coefficient of the term of greatest degree is negative.*

**15.** $24ab - 8a^2 + 40ac$

**15.** _____

**16.** $45a^2b^3 - 90ab + 15ab^2$

**16.** _____

**17.** $20x^2 + 40x^2y - 70xy^2$

**17.** _____

**18.** $2a(x - 2y) + 9b(x - 2y)$

**18.** _____

**19.** $26x^8 - 13x^{12} + 52x^{10}$

**19.** _____

**20.** $27r^2 - 54r^4 - 81r^5$

**20.** _____

**21.** $56x^2y^4 - 24xy^3 + 32xy^2$

**21.** _____

**22.** $x^2(r - 4s) + z^2(r - 4s)$

**22.** _____

Name:                                    Date:
Instructor:                              Section:

## Objective 4    Factor by grouping.

*Factor each polynomial by grouping.*

**23.**    $1 + p - q - pq$

**23.** _____

**24.**    $15 - 5x - 3y + xy$

**24.** _____

**25.**    $8x^2 + 12xy - 2xy - 3y^2$

**25.** _____

**26.**    $6x^3 + 9x^2y^2 - 2xy^3 - 3y^5$

**26.** _____

**27.**    $12x^3 - 4xy - 3x^2y^2 + y^3$

**27.** _____

**28.**    $2x^2 - 14xy + xy - 7y^2$

**28.** _____

**29.**    $3r^3 - 2r^2s + 3s^2r - 2s^3$

**29.** _____

**30.**    $2a^3 + 3a^2b + 8ab^2 + 12b^3$

**30.** _____

# Chapter 13 FACTORING AND APPLICATIONS

### 13.2   Factoring Trinomials

| Learning Objectives |
| --- |
| 1      Factor trinomials with a coefficient of 1 for the squared term. |
| 2      Factor trinomials after factoring out the greatest common factor. |

### Key Terms

Use the vocabulary terms listed below to complete each statement in exercises 1–3.

**prime polynomial    factoring    greatest common factor**

1.   _____ is the process of writing a polynomial as a product.

2.   The _____ of a polynomial is the greatest term that is a factor of all the terms in the polynomial.

3.   A _____ is a polynomial that cannot be factored using only integers.

### Objective 1   Factor trinomials with a coefficient of 1 for the squared term.

*Factor completely. If a polynomial cannot be factored, write* prime.

1.   $r^2 + r + 3$                                 1.   _____

2.   $a^2 - 10a + 21$                        2.   _____

3.   $s^2 - 4s - 32$                         3.   _____

4.   $n^2 - 16n + 64$                       4.   _____

5.   $x^2 + 11x + 18$                       5.   _____

**6.** $x^2 - 11x + 28$                **6.** _____

**7.** $x^2 - x - 2$                   **7.** _____

**8.** $x^2 + 14x - 49$                **8.** _____

**9.** $x^2 - 2x - 35$                 **9.** _____

**10.** $x^2 - 8x - 33$                **10.** _____

**11.** $x^2 - 15xy + 56y^2$           **11.** _____

**12.** $a^2 + 10ab + 21b^2$           **12.** _____

**13.** $q^2 - 4q - 12$                **13.** _____

**14.** $b^2 + 10bc + 25c^2$           **14.** _____

**15.** $a^2 - 10ab + 16b^2$           **15.** _____

**Objective 2    Factor trinomials after factoring out the greatest common factor.**

*Factor completely. If a polynomial cannot be factored, write* **prime***.*

**16.**    $3d^2 - 18d + 27$                               **16.** _____

**17.**    $2m^3 - 2m^2 - 4m$                             **17.** _____

**18.**    $2n^4 - 16n^3 + 30n^2$                         **18.** _____

**19.**    $4a^2 - 24b + 5$                               **19.** _____

**20.**    $2a^3b - 10a^2b^2 + 12ab^3$                    **20.** _____

**21.**    $3p^6 + 18p^5 + 24p^4$                         **21.** _____

**22.**    $3h^3k - 21h^2k - 54hk$                        **22.** _____

**23.**    $10k^6 + 70k^5 + 100k^4$                       **23.** _____

**24.** $3xy^2 - 24xy + 36x$

**24.** _____

**25.** $x^5 - 3x^4 + 2x^3$

**25.** _____

**26.** $a^2b - 12ab^2 + 35b^3$

**26.** _____

**27.** $2x^2y^2 - 2xy^3 - 12y^4$

**27.** _____

**28.** $2s^2t - 16st - 40t$

**28.** _____

**29.** $qr^3 - 4q^2r^2 - 21q^3r$

**29.** _____

**30.** $2x^3 - 14x^2y + 20xy^2$

**30.** _____

# Chapter 13 FACTORING AND APPLICATIONS

### 13.3    Factoring Trinomials by Grouping

**Learning Objectives**
1     Factor trinomials by grouping when the coefficient of the squared term is not 1.

### Key Terms

Use the vocabulary terms listed below to complete each statement in exercises 1–2.

> **coefficient**          **trinomial**

1.   In the term $6x^2 y$, 6 is the _____.

2.   A polynomial with three terms is a _____.

**Objective 1    Factor trinomials by grouping when the coefficient of the squared term is not 1.**

*Complete the factoring.*

1.   $2x^2 + 5x - 3 = (2x - 1)(\quad)$          1. _____

2.   $6x^2 + 19x + 10 = (3x + 2)(\quad)$          2. _____

3.   $16x^2 + 4x - 6 = (4x + 3)(\quad)$          3. _____

4.   $24y^2 - 17y + 3 = (3y - 1)(\quad)$          4. _____

*Factor each trinomial by grouping.*

5.   $8b^2 + 18b + 9$          5. _____

6.   $3x^2 + 13x + 14$          6. _____

**7.** $15a^2 + 16a + 4$

**7.** _____

**8.** $6n^2 + 11n + 4$

**8.** _____

**9.** $3b^2 + 8b + 4$

**9.** _____

**10.** $3m^2 - 5m - 12$

**10.** _____

**11.** $3p^3 + 8p^2 + 4p$

**11.** _____

**12.** $8m^2 + 26mn + 6n^2$

**12.** _____

**13.** $7a^2b + 18ab + 8b$

**13.** _____

**14.** $2s + 5st - 3st^2$

**14.** _____

**15.** $9c^2 + 24cd + 12d^2$

**15.** _____

**16.** $25a^2 + 30ab + 9b^2$

**16.** _____

**17.** $10c^2 - 29ct + 21t^2$

**17.** _____

**18.** $24s^2 - 14st - 5t^2$

**18.** _____

**19.** $12x^2 + 32xy - 35y^2$

**19.** _____

**20.** $24c^2 + 90cd - 81d^2$

**20.** _____

**21.** $6m^3 + 2m^2n - 8mn^2$

**21.** _____

**22.** $40x^2 + 18xy - 9y^2$

**22.** _____

**23.** $18f^2 + 27fg - 5g^2$                                      **23.** _____

**24.** $16p^2 + 8pq + q^2$                                        **24.** _____

**25.** $40a^3 - 82a^2 + 40a$                                      **25.** _____

**26.** $4x^2 + 32x + 55$                                          **26.** _____

**27.** $8x^2 - 4xy - 4y^2$                                        **27.** _____

**28.** $7m^2 + 3mn - 22n^2$                                       **28.** _____

**29.** $10c^2 + 39cd + 36d^2$                                     **29.** _____

**30.** $9x^3 - 30x^2y + 24xy^2$                                   **30.** _____

# Chapter 13 FACTORING AND APPLICATIONS

### 13.4    Factoring Trinomials Using FOIL

| **Learning Objectives** |
| --- |
| 1      Factor trinomials using FOIL. |

### Key Terms

Use the vocabulary terms listed below to complete each statement in exercises 1–3.

     **FOIL**          **outer product**          **inner product**

1.    The _____ of $(2y-5)(y+8)$ is $-5y$.

2.    _____ is a shortcut method for finding the product of two binomials.

3.    The _____ of $(2y-5)(y+8)$ is $16y$.

### Objective 1    Factor trinomials using FOIL.

*Factor each trinomial completely.*

1.    $10x^2 + 19x + 6$                            **1.** _____

2.    $4y^2 + 3y - 10$                            **2.** _____

3.    $2a^2 + 13a + 6$                            **3.** _____

4.    $8q^2 + 10q + 3$                            **4.** _____

5.    $8m^2 - 10m - 3$                           **5.** _____

**6.**  $14b^2 + 3b - 2$

**6.** _____

**7.**  $15q^2 - 2q - 24$

**7.** _____

**8.**  $3a^2 + 8ab + 4b^2$

**8.** _____

**9.**  $9w^2 + 12wz + 4z^2$

**9.** _____

**10.**  $10c^2 - cd - 2d^2$

**10.** _____

**11.**  $6x^2 + xy - 12y^2$

**11.** _____

**12.**  $18x^2 - 27xy + 4y^2$

**12.** _____

**13.**  $12y^2 + 11y - 15$

**13.** _____

**14.**  $3x^2 - 11x - 4$

**14.** _____

**15.** $2p^2 + 11p + 5$

**15.** _____

**16.** $6y^2 + y - 1$

**16.** _____

**17.** $9y^2 - 16y - 4$

**17.** _____

**18.** $3p^2 + 17p + 10$

**18.** _____

**19.** $9r^2 + 12r - 5$

**19.** _____

**20.** $7x^2 + 27x - 4$

**20.** _____

**21.** $4c^2 + 14cd - 8d^2$

**21.** _____

**22.** $2x^4 + 5x^3 - 12x^2$

**22.** _____

**23.** $27r^2 + 6rt - 8t^2$

**23.** _____

**24.** $6x^4y^2 + x^2y - 15$

**24.** _____

**25.** $28c^2 + 23cd - 15d^2$

**25.** _____

**26.** $8x^3 - 10x^2y + 3xy^2$

**26.** _____

**27.** $6n^2 + 13ns - 63s^2$

**27.** _____

**28.** $-30a^4b + 3a^3b + 6a^2b$

**28.** _____

**29.** $12a^3 + 26a^2b + 12ab^2$

**29.** _____

**30.** $2y^5z^2 - 5y^4z^3 - 3y^3z^4$

**30.** _____

# Chapter 13 FACTORING AND APPLICATIONS

### 13.5   Special Factoring Techniques

| Learning Objectives |
| --- |
| 1       Factor a difference of squares. |
| 2       Factor a perfect square trinomial. |
| 3       Factor a difference of cubes. |
| 4       Factor a sum of cubes |

### Key Terms

Use the vocabulary terms listed below to complete each statement in exercises 1–2.

**perfect square trinomial      difference of squares**

1.   A _____ is a binomial that can be factored as the product of the sum and difference of two terms.

2.   A _____ is a trinomial that can be factored as the square of a binomial.

### Objective 1   Factor a difference of squares.

*Factor each binomial completely. If a binomial cannot be factored, write* prime.

1.   $25a^2 - 36$                                     1. _____

2.   $x^2 + 16$                                        2. _____

3.   $9j^2 - \frac{16}{49}$                            3. _____

4.   $121m^2 - 9n^2$                                   4. _____

5.   $16y^4 - 81$                                      5. _____

**6.**  $q^2 - (2r+3)^2$

**6.** _____

**7.**  $m^4 n^2 - m^2$

**7.** _____

**8.**  $(r-s)^2 - (r+s)^2$

**8.** _____

**Objective 2    Factor a perfect square trinomial.**

*Factor each polynomial completely.*

**9.**  $4w^2 + 12w + 9$

**9.** _____

**10.**  $z^2 - \frac{4}{3}z + \frac{4}{9}$

**10.** _____

**11.**  $16q^2 - 40q + 25$

**11.** _____

**12.**  $64p^4 + 48p^2 q^2 + 9q^4$

**12.** _____

**13.**  $100p^2 - \frac{25}{2}pr + \frac{25}{64}r^2$

**13.** _____

**14.**  $9m^2 + .6m + .01$

**14.** _____

**15.**    $(p-q)^2 - 20(p-q) + 100$                    **15.** _____

**16.**    $(m-n)^2 - 12(m-n) + 36$                    **16.** _____

**Objective 3    Factor a difference of cubes.**

*Factor.*

**17.**    $x^3 - y^3$                    **17.** _____

**18.**    $8r^3 - 27s^3$                    **18.** _____

**19.**    $216m^3 - 125p^6$                    **19.** _____

**20.**    $8a^3 - 125b^3$                    **20.** _____

**21.**    $216x^3 - 8y^3$                    **21.** _____

**22.**    $(m+n)^3 - (m-n)^3$                    **22.** _____

**23.**    $x^3 - (x-1)^3$                    **23.** _____

## Objective 4    Factor a sum of cubes.

*Factor.*

**24.**    $x^3 + y^3$                        **24.** _____

**25.**    $27r^3 + 8s^3$                     **25.** _____

**26.**    $8a^3 + 64b^3$                     **26.** _____

**27.**    $125p^3 + q^3$                     **27.** _____

**28.**    $64x^3 + 343y^3$                   **28.** _____

**29.**    $(x-y)^3 + (x+y)^3$                **29.** _____

**30.**    $t^3 + (t+2)^3$                    **30.** _____

# Chapter 13 FACTORING AND APPLICATIONS

### 13.6    A General Approach to Factoring

| **Learning Objectives** | |
|---|---|
| 1 | Factor out any common factor. |
| 2 | Factor binomials. |
| 3 | Factor trinomials. |
| 4 | Factor polynomials with more than three terms. |

### Key Terms

Use the vocabulary terms listed below to complete each statement in exercises 1–2.

**FOIL            factoring by grouping**

1.    When there are more than three terms in a polynomial, use a process called
_____ to factor the polynomial.

2.    _____ is a shortcut method for finding the product of two
binomials.

### Objective 1    Factor out any common factor.

*Factor completely.*

1.    $-12x^2 - 6x$                              1. _____

2.    $12a^2b^2 + 3a^2b - 9ab^2$               2. _____

3.    $5r^2t - 10rt + 5rt^2$                    3. _____

**4.**  $12x^4 - 8x^2 + 20x$          **4.** _____

**5.**  $3a^2 + 6a(x - y)$          **5.** _____

**6.**  $(x+1)(2x+3) - (x+1)$          **6.** _____

**7.**  $2m(m-n) - (m+n)(m-n)$          **7.** _____

**Objective 2    Factor binomials.**

*Factor completely.*

**8.**  $16s^4 - r^4$          **8.** _____

**9.**  $2x^3y^4 - 72xy^2$          **9.** _____

**10.**   $128x^3 - 2y^3$

**10.** _____

**11.**   $(r+s)^3 + 8$

**11.** _____

**12.**   $(a-b)^2 - (a+b)^2$

**12.** _____

**13.**   $y^6 + 1$

**13.** _____

**14.**   $32 - 2(x-y)^2$

**14.** _____

**15.**   $(3a-1)^2 - y^6$

**15.** _____

## Objective 3   Factor trinomials.

*Factor completely.*

**16.**    $4x^2 + 12xy + 9y^2$                            **16.** _____

**17.**    $2a^2 - 17a + 30$                               **17.** _____

**18.**    $4y^4 + 8y^2 - 45$                              **18.** _____

**19.**    $25x^2 - 5xy - 2y^2$                            **19.** _____

**20.**    $3(a+1)^2 + 19(a+1) - 14$                       **20.** _____

**21.**    $12m^2 + 11m - 5$                               **21.** _____

**22.**    $3k + 42k^2 + 147k^3$                          22. _____

**23.**    $4b^2(b+2) - 3b(b+2) - (b+2)$                  23. _____

**Objective 4    Factor polynomials with more than three terms.**

*Factor completely.*

**24.**    $14w^2 + 6wx - 35wx - 15x^2$                    24. _____

**25.**    $bx - by - ay + ax$                            25. _____

**26.**    $x^2 + 7xy + 2x + 14y$                         26. _____

**27.**    $x^3 - 21 - 3x^2 + 7x$                         27. _____

**28.**   $3x^2 - 2x + 6x - 4$

**28.** _____

**29.**   $a^2 - 6ab + 9b^2 - 25$

**29.** _____

**30.**   $r^3 - s^3 - sr^2 + rs^2$

**30.** _____

# Chapter 13 FACTORING AND APPLICATIONS

### 13.7    Solving Quadratic Equations by Factoring

| **Learning Objectives** |
| --- |
| 1     Solve quadratic equations by factoring. |
| 2     Solve other equations by factoring. |

### Key Terms

Use the vocabulary terms listed below to complete each statement in exercises 1–2.

    **quadratic equation**       **standard form**

1.    An equation written in the form $ax^2 + bx + c = 0$ is written in the
      _____ of a quadratic equation.

2.    An equation that can written in the form $ax^2 + bx + c = 0$, with $a \neq 0$, is a
      _____.

### Objective 1    Solve quadratic equations by factoring.

*Solve each equation and check your solutions.*

1.    $x^2 + 7x + 10 = 0$                       **1.** _____

2.    $3x^2 + 7x + 2 = 0$                   **2.** _____

3.    $b^2 - 49 = 0$                           **3.** _____

**4.** $2x^2 - 3x - 20 = 0$

**4.** _____

**5.** $x^2 - 2x - 63 = 0$

**5.** _____

**6.** $8r^2 = 24r$

**6.** _____

**7.** $3x^2 - 7x - 6 = 0$

**7.** _____

**8.** $3 - 5x = 8x^2$

**8.** _____

**9.** $9x^2 + 12x + 4 = 0$

**9.** _____

**10.** $25x^2 = 20x$

**10.** _____

**11.**    $9y^2 = 16$                       **11.** _____

**12.**    $12x^2 + 7x - 12 = 0$          **12.** _____

**13.**    $14x^2 - 17x - 6 = 0$          **13.** _____

**14.**    $c(5c + 17) = 12$              **14.** _____

**15.**    $3x(x + 3) = (x + 2)^2 - 1$     **15.** _____

## Objective 2    Solve other equations by factoring.

*Solve each equation and check your solutions.*

**16.**    $3x(x + 7)(x - 2) = 0$          **16.** _____

**17.** $x\left(2x^2 - 7x - 15\right) = 0$

**17.** _____

**18.** $z\left(4z^2 - 9\right) = 0$

**18.** _____

**19.** $z^3 - 49z = 0$

**19.** _____

**20.** $25a = a^3$

**20.** _____

**21.** $x^3 + 2x^2 - 8x = 0$

**21.** _____

**22.** $2m^3 + m^2 - 6m = 0$

**22.** _____

**23.** $\left(4x^2 - 9\right)(x - 2) = 0$

**24.** $z^4 + 8z^3 - 9z^2 = 0$

**25.** $3z^3 + z^2 - 4z = 0$

**26.** $(x + 4)\left(x^2 + 7x + 10\right) = 0$

**27.** $\left(y^2 - 5y + 6\right)\left(y^2 - 36\right) = 0$

**28.** $15x^2 = x^3 + 56x$

**29.**    $(y - 7)\left(2y^2 + 7y - 15\right) = 0$                **29.** _____

**30.**    $\left(x - \frac{3}{2}\right)\left(2x^2 + 11x + 15\right) = 0$                **30.** _____

# Chapter 13 FACTORING AND APPLICATIONS

## 13.8    Applications of Quadratic Equations

| **Learning Objectives** | |
|---|---|
| 1 | Solve problems about geometric figures. |
| 2 | Solve problems about consecutive integers. |
| 3 | Solve problems using the Pythagorean formula. |
| 4 | Solve problems using given quadratic models. |

### Key Terms

Use the vocabulary terms listed below to complete each statement in exercises 1–2.

**hypotenuse**         **legs**

1.    In a right triangle, the sides that form the right angle are the _____.

2.    The longest side of a right triangle is the _____.

### Objective 1    Solve problems about geometric figures.

*Solve each problem. Check your answers to be sure they are reasonable.*

1.    A book is three times as long as it is wide. Find the length and width of the book in inches if its area is numerically 128 more than its perimeter.

     1. width_____

        length _____

2.    The length of a rectangle is three times its width. If the width were increased by 4 and the length remained the same, the resulting rectangle would have an area of 231 square inches. Find the dimensions of the original rectangle.

     2. width_____

        length _____

**3.** Two rectangles with different dimensions have the same area. The length of the first rectangle is three times its width. The length of the second rectangle is 4 meters more than the width of the first rectangle, and its width is 2 meters more than the width of the first rectangle. Find the lengths and widths of the two rectangles.

**3.** Rectangle 1:

width _____

length _____

Rectangle 2:

width _____

length _____

**4.** Each side of one square is 1 meter less than twice the length of each side of a second square. If the difference between the areas of the two squares is 16 meters, find the lengths of the sides of the two rectangles.

**4.** square 1 _____

square 2 _____

**5.** The area of a triangle is 42 square centimeters. The base is 2 centimeters less than twice the height. Find the base and height of the triangle.

**5.** base _____

height _____

**6.** The volume of a box is 192 cubic feet. If the length of the box is 8 feet and the width is 2 feet more than the height, find the height and width of the box.

**6.** height _____

width _____

**7.** Mr. Fixxall is building a box which will have a volume of 60 cubic meters. The height of the box will be 4 meters, and the length will be 2 meters more than the width. Find the width and length of the box.

**7.** width _____

length _____

**Objective 2   Solve problems about consecutive integers.**

*Solve each problem.*

**8.** Find two consecutive integers such that the sum of the squares of the two integers is 3 more than the opposite (additive inverse) of the smaller integer.

**8.** _____

**9.** If the square of the sum of two consecutive integers is reduced by twice their product, the result is 25. Find the integers.

**9.** _____

**10.** Find all possible pairs of consecutive odd integers whose sum is equal to their product decreased by 47.

**10.** _____

**11.** Find two consecutive positive even integers whose product is 168.

**11.** _____

**12.** Find two consecutive positive even integers whose product is six more than three times its sum.

**12.** _____

**13.** The product of two consecutive even positive integers is 10 more than seven times the larger. Find the integers.

**13.** _____

**14.** Find three consecutive positive odd integers such that four times the sum of all three equals 13 more than the product of the smaller two.

**14.** _____

**Objective 3    Solve problems using the Pythagorean formula.**

*Solve each problem.*

**15.** The hypotenuse of a right triangle is 4 inches longer than the shorter leg. The longer leg is 4 inches shorter than twice the shorter leg. Find the lengths of the three sides.

**15.** _____

**16.**   A field is in the shape of a right triangle. The shorter leg measures 45 meters. The hypotenuse measures 45 meters less than twice the longer the leg. Find the dimensions of the lot.

**16.** _____

**17.**   A train and a car leave a station at the same time, the train traveling due north and the car traveling west. When they are 100 miles apart, the train has traveled 20 miles farther than the car. Find the distance each has traveled.

**17.** car _____

train _____

**18.**   Penny and Carla started biking from the same corner. Penny biked east and Carla biked south. When they were 26 miles apart, Carla had biked 14 miles further than Penny. Find the distance each biked.

**18.** Penny _____

Carla _____

**19.** Two trains leave New York City at the same time. One train travels due north and the other travels due east. When they are 75 miles apart, the train going north has gone 30 miles less than twice the distance traveled by the train going east. Find the distance traveled by the train going east.

**19.** _____

**20.** Mark is standing directly beneath a kite attached to a string which Nina is holding, with her hand touching the ground. The height of the kite at that instant is 12 feet less than twice the distance between Mark and Nina. The length of the kite string is 12 feet more than the distance between Mark and Nina. Find the length of the kite string.

**20.** _____

**21.** Two ships left a dock at the same time. When they were 25 miles apart, the ship that sailed due south had gone 10 miles less than twice the distance traveled by the ship that sailed due west. Find the distance traveled by the ship that sailed due south.

**21.** _____

**22.**     A ladder is leaning against a building. The distance from the bottom of the ladder to the building is 8 feet less than the length of the ladder. How high up the side of the building is the top of the ladder if that distance is 4 feet less than the length of the ladder?

**22.** _____

**Objective 4**     **Solve problems using given quadratic models.**

*Solve each problem.*

**23.**     A ball is dropped from the roof of a 19.6 meter high building. Its height $h$ (in meters) $t$ seconds later is given by the equation $h = -4.9t^2 + 19.6$.

    (a)   After how many seconds is the height 18.375 meters?

    (b)   After how many seconds is the height 14.7 meters?

    (c)   After how many seconds does the ball hit the ground?

**23. a.** _____

     **b.** _____

     **c.** _____

**24.** If an object is propelled upward from a height of 16 feet with an initial velocity of 48 feet per second, its height $h$ (in feet) $t$ seconds later is given by the equation $h = -16t^2 + 48t + 16$.

(a)  After how many seconds is the height 52 feet?

(b)  After how many seconds is the height 48 feet?

**24. a.** _____

**b.** _____

**25.** The total cost of a product can be modeled by the equation $C = 400 - 100x + x^2$ where $x$ represents the number of items produced. How many items can be produced at a cost of \$1500?

**25.** _____

**26.** Jeff threw a stone straight upward at 46 feet per second from a dock 6 feet above a lake. The height of the stone above the lake $t$ seconds after it is thrown is given by $h = -16t^2 + 46t + 6$. How long will it take for the stone to reach a height of 39 feet?

**26.** _____

27.  A company determines that its daily revenue $R$ (in dollars) for selling $x$ items is modeled by the equation $R = x(150 - x)$. How many items must be sold for its revenue to be $4400?

27. _____

28.  If a ball is batted at an angle of 35°, the distance that the ball travels is given approximately by $D = 0.029v^2 + 0.021v - 1,$ where $v$ is the bat speed in miles per hour and $D$ is the distance traveled in feet. Find the distance a batted ball will travel if the ball is batted with a velocity of 90 miles per hour. Round your answer to the nearest whole number.

28. _____

**29.** Altitude affects the distance a batted ball travels. The data shown below can be modeled by the equation $D = -0.000000234a^2 + .0069a + 400$, where $D$ is the distance the ball travels in feet and $a$ is the altitude above sea level in feet. Use the model to find how far a ball will travel in Wrigley Field where the altitude is approximately 580 feet. Round your answer to the nearest whole number.
(Source: Robert K. Adair, *The Physics of Baseball*, (Harper Perennial, 1994), 18–19)

**29.** _____

| Stadium | Altitude (ft) | Distance (ft) |
|---|---|---|
| Yankee Stadium (New York) | 0 | 400 |
| Kauffman Stadium (Kansas City) | 740 | 405 |
| Turner Field (Atlanta) | 1050 | 407 |
| Coors Field (Denver) | 5280 | 430 |

**30.** The unemployment rate in a certain community can be modeled by the equation $y = 0.0248x^2 - 0.4810x + 7.8543,$ where $y$ is the unemployment rate (percent) and $x$ is the month ($x = 1$ represents January, $x = 2$ represents February, etc.) Use the model to find the unemployment rate in August. Round your answer to the nearest tenth.

**30.** _____

# Chapter 14 RATIONAL EXPRESSIONS AND APPLICATIONS

### 14.1    The Fundamental Property of Rational Expressions

| Learning Objectives | |
| --- | --- |
| 1 | Find the values of the variable for which a rational expression is undefined. |
| 2 | Find the numerical value of a rational expression. |
| 3 | Write rational expressions in lowest terms. |
| 4 | Recognize equivalent forms of rational expressions. |

### Key Terms

Use the vocabulary terms listed below to complete each statement in exercises 1–2.

**rational expression            lowest terms**

1.    The quotient of two polynomials with denominator not 0 is called a

_____.

2.    A rational expression is written in _____ if the
greatest common factor of its numerator and denominator is 1.

### Objective 1    Find the values of the variable for which a rational expression is undefined.

*Find any value(s) of the variable for which each rational expression is undefined. Write answers with ≠ .*

1.    $\dfrac{4x^2}{x+7}$

1. _____

2.    $\dfrac{x-4}{4x^2-16x}$

2. _____

3.    $\dfrac{5x}{x^2-25}$

3. _____

**4.** $\dfrac{2x^2}{x^2+4}$

**4.** _____

**5.** $\dfrac{3z}{z^2-6z+9}$

**5.** _____

**6.** $\dfrac{2y-5}{2y^2+4y-16}$

**6.** _____

**7.** $\dfrac{y+2}{y^4-16}$

**7.** _____

**Objective 2    Find the numerical value of a rational expression.**

*Find the numerical value of each expression when (a) $x=4$ and (b) $x=-3$.*

**8.** $\dfrac{3x^2-2x}{2x}$

**8. (a)** _____

**(b)** _____

**9.** $\dfrac{-3x+1}{2x+1}$

**9. (a)** _____

**(b)** _____

**10.** $\dfrac{2x+5}{4x^2-25}$

**10. (a)** _____

**(b)** _____

**11.**  $\dfrac{3x}{(3x-2)^2}$

**11. (a)** _____

**(b)** _____

**12.**  $\dfrac{2x^2-4}{x^2-2}$

**12. (a)** _____

**(b)** _____

**13.**  $\dfrac{2x-5}{2+x-x^2}$

**13. (a)** _____

**(b)** _____

**14.**  $\dfrac{-2x^2}{2x^2-x+2}$

**14. (a)** _____

**(b)** _____

**Objective 3    Write rational expressions in lowest terms.**

*Write each rational expression in lowest terms. Assume that no values of any variable make any denominator zero.*

**15.**  $\dfrac{15ab^3c^9}{-24ab^2c^{10}}$

**15.** _____

**16.** $\dfrac{16-x^2}{2x-8}$

**16.** _____

**17.** $\dfrac{12k^3+12k^2}{3k^2+3k}$

**17.** _____

**18.** $\dfrac{16r^3-8r^2}{8r^2-4r^3}$

**18.** _____

**19.** $\dfrac{9x^2-9x-108}{2x-8}$

**19.** _____

**20.** $\dfrac{3y^2-13y-10}{2y^2-9y-5}$

**20.** _____

**21.** $\dfrac{6r^2-7rs-10s^2}{r^2+3rs-10s^2}$

**21.** _____

**22.** $\dfrac{5x^2 - 17xy - 12y^2}{x^2 - 7xy + 12y^2}$

22. _____

**23.** $\dfrac{vw - 5v + 3w - 15}{vw - 5v - 2w + 10}$

23. _____

**24.** $\dfrac{2r^3 + 2rs^2 + 3r^2s + 3s^3}{4r^3 + 4rs^2 - r^2s - s^3}$

24. _____

**Objective 4   Recognize equivalent forms of rational expressions.**

*Write four equivalent forms of the following rational expressions. Assume that no values of any variable make any denominator zero.*

**25.** $-\dfrac{4x + 5}{3 - 6x}$

25. _____

**26.** $\dfrac{2p - 1}{1 - 4p}$

26. _____

**27.** $-\dfrac{2x - 3}{x + 2}$

27. _____

**28.**  $\dfrac{4x+1}{5x-3}$

**28.** _____

**29.**  $-\dfrac{2x-1}{3x+5}$

**29.** _____

**30.**  $-\dfrac{2x+6}{x-5}$

**30.** _____

# Chapter 14 RATIONAL EXPRESSIONS AND APPLICATIONS

### 14.2    Multiplying and Dividing Rational Expressions

| Learning Objectives |
| --- |
| 1    Multiply rational expressions. |
| 2    Find reciprocals. |
| 3    Divide rational expressions. |

### Key Terms

Use the vocabulary terms listed below to complete each statement in exercises 1–3.

**rational expression**      **reciprocal**     **lowest terms**

1.    The _____ of the expression $\dfrac{4x-5}{x+2}$ is $\dfrac{x+2}{4x-5}$.

2.    A _____ is the quotient of two polynomials with denominator not 0.

3.    A rational expression is written in _____ when the numerator and denominator have no common terms.

### Objective 1    Multiply rational expressions.

*Multiply. Write each answer in lowest terms.*

1.    $\dfrac{8m^4 n^3}{3} \cdot \dfrac{5}{4mn^2}$                   1. _____

2.    $\dfrac{6}{9y+36} \cdot \dfrac{4y+16}{9}$             2. _____

3.    $\dfrac{r-s}{12} \cdot \dfrac{8}{s-r}$                  3. _____

**4.**   $\dfrac{4r+4p}{8z^2} \cdot \dfrac{36z^6}{r^2+rp}$

**4.** _____

**5.**   $\dfrac{2x+1}{16-x^2} \cdot \dfrac{x-4}{4x+2}$

**5.** _____

**6.**   $\dfrac{m^2-16}{m-3} \cdot \dfrac{9-m^2}{4-m}$

**6.** _____

**7.**   $\dfrac{x^2-4}{2x^2-2} \cdot \dfrac{x-x^2}{2x^2+4x}$

**7.** _____

**8.**   $\dfrac{3x+12}{6x-30} \cdot \dfrac{x^2-x-20}{x^2-16}$

**8.** _____

**9.** $\dfrac{x^2+x-12}{x^2+7x+10} \cdot \dfrac{x^2+3x-10}{x^2+2x-8}$

9. _____

**10.** $\dfrac{2x^2+5x-12}{x^2-2x-24} \cdot \dfrac{x^2-9x+18}{9-4x^2}$

10. _____

**11.** $\dfrac{x^2+10x+21}{x^2+14x+49} \cdot \dfrac{x^2+12x+35}{x^2-6x-27}$

11. _____

**12.** $\dfrac{3m^2-m-10}{2m^2-7m-4} \cdot \dfrac{4m^2-1}{6m^2+7m-5}$

12. _____

**Objective 2  Find reciprocals.**

*Write the reciprocal of each rational expression.*

**13.** $\dfrac{7}{2y}$

13. _____

**14.** $\dfrac{5a+7}{9b^2c^3}$

**14.** _____

**15.** $4r^2 + 2r + 3$

**15.** _____

**16.** $\dfrac{3x+4y}{5x-2y}$

**16.** _____

**17.** $\dfrac{x^2 - 2x + 3}{2x^2 + 9}$

**17.** _____

**18.** $\dfrac{4}{2x^3 + 5x^2 + x - 7}$

**18.** _____

## Objective 3    Divide rational expressions.

*Divide. Write each answer in lowest terms.*

**19.** $\dfrac{b-7}{16} \div \dfrac{7-b}{8}$

**19.** _____

**20.** $\dfrac{2x+2y}{8z} \div \dfrac{x^2\left(x^2 - y^2\right)}{24}$

**20.** _____

**21.**    $\dfrac{4m-12}{2m+10} \div \dfrac{m^2-9}{m^2-25}$

21. _____

**22.**    $\dfrac{m^2+2mn+n^2}{m^2+m} \div \dfrac{m^2-n^2}{m^2-1}$

22. _____

**23.**    $\dfrac{9a^2-1}{9a^2-6a+1} \div \dfrac{3a^2-11a-4}{12a^2+5a-3}$

23. _____

**24.**    $\dfrac{2z^2-11z-21}{z^2-5z-14} \div \dfrac{4z^2-9}{z^2-6z-16}$

24. _____

**25.**    $\dfrac{y^2+yz-12z^2}{y^2+yz-20z^2} \div \dfrac{y^2+9yz+20z^2}{y^2-yz-30z^2}$

25. _____

**26.** $\dfrac{27 - 3k^2}{3k^2 + 8k - 3} \div \dfrac{k^2 - 6k + 9}{6k^2 - 19k + 3}$

**26.** _____

**27.** $\dfrac{y^2 + 7y + 10}{3y + 6} \div \dfrac{y^2 + 2y - 15}{4y - 4}$

**27.** _____

**28.** $\dfrac{2y^2 - 21y - 11}{2 - 8y^2} \div \dfrac{y^2 - 12y + 11}{4y^2 + 14y - 8}$

**28.** _____

**29.** $\dfrac{2k^2 + 5k - 12}{2k^2 + k - 3} \div \dfrac{k^2 + 8x + 16}{2k^2 + 11k + 12}$

**29.** _____

**30.** $\dfrac{z^4 + 2z^3 + z^2}{z^5 - 4z^3} \div \dfrac{9z + 9}{6z + 12}$

**30.** _____

# Chapter 14 RATIONAL EXPRESSIONS AND APPLICATIONS

### 14.3    Least Common Denominators

| Learning Objectives |
| --- |
| 1     Find the least common denominator for a list of fractions. |
| 2     Write equivalent rational expressions. |

### Key Terms

Use the vocabulary terms listed below to complete each statement in exercises 1–2.

**least common denominator          equivalent expressions**

1.    $\dfrac{24x-8}{9x^2-1}$ and $\dfrac{8}{3x+1}$ are _____.

2.    The simplest expression that is divisible by all denominators is called the

_____.

### Objective 1    Find the least common denominator for a list of fractions.

*Find the least common denominator for each list of rational expressions.*

1.    $\dfrac{7}{16r^3}, \dfrac{9}{12r^4}$                                    1. _____

2.    $\dfrac{5}{8ab^2}, \dfrac{7}{6a^2b}$                                    2. _____

3.    $\dfrac{13}{36b^4}, \dfrac{17}{27b^2}$                                    3. _____

4.    $\dfrac{4}{5r-25}, \dfrac{7}{15r^3}$                                    4. _____

5.    $\dfrac{15}{7t-28}, \dfrac{21}{6t-24}$                                    5. _____

**6.** $\dfrac{7}{x-y}, \dfrac{3}{y-x}$

**6.** _____

**7.** $\dfrac{4}{a^2-b^2}, \dfrac{8}{b^2-a^2}$

**7.** _____

**8.** $\dfrac{3m}{2m^2+9m-5}, \dfrac{4}{m^2+5m}$

**8.** _____

**9.** $\dfrac{-7}{a^2-2a}, \dfrac{3a}{2a^2+a-10}$

**9.** _____

**10.** $\dfrac{8}{w^3-9w}, \dfrac{4w}{w^2+w-6}$

**10.** _____

**11.** $\dfrac{7t}{t^2+6t+8}, \dfrac{4t}{t^2+t-12}$

**11.** _____

**12.** $\dfrac{3z+1}{z^4+2z^3-8z^2}, \dfrac{5z+2}{z^3+8z^2+16z}$

**12.** _____

**13.** $\dfrac{11q-3}{2q^2-q-10}, \dfrac{21-q}{2q^2-9q+10}$

**13.** _____

**14.** $\dfrac{17r}{9r^2-6r-8}, \dfrac{-13r}{9r^2-9r-4}$

**14.** _____

**15.** $\dfrac{m+2}{m^3-2m^2}, \dfrac{3-m}{m^2+5n-14}$

**15.** _____

**Objective 2    Write equivalent rational expressions.**

*Rewrite each rational expression with the indicated denominator. Give the numerator of the new fraction.*

**16.** $\dfrac{7m}{8n} = \dfrac{?}{24n^6}$

**16.** _____

**17.** $\dfrac{3}{5w} = \dfrac{?}{40w^2}$

**17.** _____

**18.** $\dfrac{5a}{8a-3} = \dfrac{?}{6-16a}$

**18.** _____

**19.** $\dfrac{2}{3c-5} = \dfrac{?}{12c-20}$

**19.** _____

**20.** $\dfrac{5a}{7a-14} = \dfrac{?}{28a^2-56a}$

**21.** $\dfrac{11a+1}{2a-6} = \dfrac{?}{8a-24}$

**22.** $\dfrac{-3y}{4y+12} = \dfrac{?}{4(y+3)^2}$

**23.** $\dfrac{8z}{3z+3} = \dfrac{?}{12z^2+15z+3}$

**24.** $\dfrac{3}{5r-10} = \dfrac{?}{50r^2-100r}$

**25.** $\dfrac{9}{y^2-4} = \dfrac{?}{(y+2)^2(y-2)}$

**26.** $\dfrac{5}{2r+8} = \dfrac{?}{2(r+4)(r^2+2r-8)}$

**27.**  $\dfrac{3x+1}{x^2-4}=\dfrac{?}{2x^2-8}$

**27.** _____

**28.**  $\dfrac{2}{7p-35}=\dfrac{?}{14p^3-70p^2}$

**28.** _____

**29.**  $\dfrac{3}{k^2+3k}=\dfrac{?}{k^3+10k^2+21k}$

**29.** _____

**30.**  $\dfrac{r+1}{r^2+2r}=\dfrac{?}{2r^3+3r^2-2r}$

**30.** _____

# Chapter 14 RATIONAL EXPRESSIONS AND APPLICATIONS

### 14.4    Adding and Subtracting Rational Expressions

| **Learning Objectives** |
| --- |
| 1    Add rational expressions having the same denominator. |
| 2    Add rational expressions having different denominators. |
| 3    Subtract rational expressions. |

### Key Terms

Use the vocabulary terms listed below to complete each statement in exercises 1–2.

**least common multiple**          **greatest common factor**

1.    The _____ of $2m^2 - 5m - 3$ and $2m - 6$ is $m - 3$.

2.    The _____ of $2m^2 - 5m - 3$ and $2m - 6$ is $2(m - 3)(2m + 1)$.

### Objective 1    Add rational expressions having the same denominator.

*Add. Write each answer in lowest terms.*

1.    $\dfrac{5}{3w^2} + \dfrac{7}{3w^2}$

1. _____

2.    $\dfrac{5t + 4}{2t + 1} + \dfrac{4t + 2}{2t + 1}$

2. _____

3.    $\dfrac{b}{b^2 - 4} + \dfrac{2}{b^2 - 4}$

3. _____

4.    $\dfrac{2x + 3}{x^2 + 3x - 10} + \dfrac{2 - x}{x^2 + 3x - 10}$

4. _____

5. $\dfrac{2y-5}{2y^2-5y-3}+\dfrac{2-y}{2y^2-5y-3}$ 

5. _____

6. $\dfrac{6v}{6v^2+13vw+6w^2}+\dfrac{4w}{6v^2+13vw+6w^2}$ 

6. _____

**Objective 2    Add rational expressions having different denominators.**

*Add. Write each answer in lowest terms.*

7. $\dfrac{7}{x-5}+\dfrac{4}{x+5}$ 

7. _____

8. $\dfrac{3z-2}{5z+20}+\dfrac{2z+1}{3z+12}$ 

8. _____

9. $\dfrac{2m+3}{m^2-3m-4}+\dfrac{3m-2}{m^2-16}$ 

9. _____

10. $\dfrac{3p}{4p^2-9}+\dfrac{4}{6p+9}$ 

10. _____

**11.** $\dfrac{2s+3}{3s^2-14s+8}+\dfrac{4s+5}{2s^2-5s-12}$

**11.** _____

**12.** $\dfrac{5p-2}{2p^2+9p+9}+\dfrac{p+7}{6p^2+13p+6}$

**12.** _____

**13.** $\dfrac{3z}{z^2-4}+\dfrac{4z-3}{z^2-4z+4}$

**13.** _____

**14.** $\dfrac{1-3x}{4x^2-1}+\dfrac{3x-5}{2x^2+5x+2}$

**14.** _____

**15.** $\dfrac{3}{6x^2+x-2}+\dfrac{-1}{3x^2+8x+4}$

**15.** _____

**16.** $\dfrac{3}{9b^2-16}+\dfrac{2}{3b^2+2b-8}$

**16.** _____

**17.** $\dfrac{4}{2c^2 + c - 3} + \dfrac{c}{2c^2 - 5c - 12}$                  **17.** _____

**18.** $\dfrac{4z}{z^2 + 6z + 8} + \dfrac{2z - 1}{z^2 + 5z + 6}$                  **18.** _____

## Objective 3    Subtract rational expressions.

*Subtract. Write each answer in lowest terms.*

**19.** $\dfrac{z + 2}{z - 2} - \dfrac{z - 2}{z + 2}$                  **19.** _____

**20.** $\dfrac{6}{x - y} - \dfrac{4 + y}{y - x}$                  **20.** _____

**21.** $\dfrac{-4}{x^2 - 4} - \dfrac{3}{4 - 2x}$                  **21.** _____

**22.**   $\dfrac{1}{9m-3} - \dfrac{m-2}{3m^2+11m-4}$

**23.**   $\dfrac{2}{4d^2+8d+3} - \dfrac{1}{2d+1}$

**24.**   $\dfrac{3}{n^2-16} - \dfrac{6n}{n^2+8n+16}$

**25.**   $\dfrac{m}{m^2-4} - \dfrac{1-m}{m^2+4m+4}$

**26.**   $\dfrac{6}{2q^2+5q+2} - \dfrac{5}{2q^2-3q-2}$

**27.**   $\dfrac{5}{2c^2+7c+6} - \dfrac{3}{2c^2-c-6}$

**28.** $\dfrac{4y}{y^2+4y+3} - \dfrac{3y+1}{y^2-y-2}$

**28.** _____

**29.** $\dfrac{4x-1}{2x^2+5x-3} - \dfrac{x+3}{6x^2+x-2}$

**29.** _____

**30.** $\dfrac{6z}{(z+1)^2} - \dfrac{2z+3}{z^2-1}$

**30.** _____

# Chapter 14 RATIONAL EXPRESSIONS AND APPLICATIONS

### 14.5    Complex Fractions

**Learning Objectives**
1    Simplify a complex fraction by writing it as a division problem (Method 1).
2    Simplify a complex fraction by multiplying numerator and denominator by the least common denominator (Method 2).

### Key Terms

Use the vocabulary terms listed below to complete each statement in exercises 1–2.

    **complex fraction**    **LCD**

1.    A _____ is a rational expression with one or more fractions in the numerator, denominator, or both.

2.    To simplify a complex fraction, multiply the numerator and denominator by the _____ of all the fractions within the complex fraction.

### Objective 1    Simplify a complex fraction by writing it as a division problem (Method 1).

*Simplify each complex fraction by writing it as a division problem.*

1.    $\dfrac{-\dfrac{3}{5}}{\dfrac{9}{10}}$

    1. _____

2.    $\dfrac{\dfrac{3}{4}-\dfrac{1}{2}}{\dfrac{1}{4}+\dfrac{5}{8}}$

    2. _____

3.    $\dfrac{\dfrac{49m^3}{18n^5}}{\dfrac{21m}{27n^2}}$

    3. _____

**4.** $\dfrac{\dfrac{r-s}{12}}{\dfrac{r^2-s^2}{6}}$

**5.** $\dfrac{2-\dfrac{1}{y-2}}{3-\dfrac{2}{y-2}}$

**6.** $\dfrac{3-\dfrac{5}{m}}{\dfrac{2}{m}+2}$

**7.** $\dfrac{\dfrac{p}{2}-\dfrac{1}{3}}{\dfrac{p}{3}+\dfrac{1}{6}}$

**8.** $\dfrac{\dfrac{4}{z}+2}{\dfrac{1+z}{2}}$

**9.** $\dfrac{3+\dfrac{4}{s}}{2s+\dfrac{2}{3}}$

**10.** $\dfrac{\dfrac{4}{p} - 2p}{\dfrac{3 - p^2}{6}}$

**11.** $\dfrac{\dfrac{a+2}{a-2}}{\dfrac{1}{a^2-4}}$

**12.** $\dfrac{\dfrac{2}{a+2} - 4}{\dfrac{1}{a+2} - 3}$

**13.** $\dfrac{\dfrac{3}{w-4} - \dfrac{3}{w+4}}{\dfrac{1}{w+4} + \dfrac{1}{w^2-16}}$

**14.** $\dfrac{\dfrac{5}{rs^2} - \dfrac{2}{rs}}{\dfrac{3}{rs} - \dfrac{4}{r^2 s}}$

**15.** $\dfrac{\dfrac{3a+4}{a}}{\dfrac{1}{a} + \dfrac{2}{5}}$

**10.** _____

**11.** _____

**12.** _____

**13.** _____

**14.** _____

**15.** _____

**Objective 2    Simplify a complex fraction by multiplying numerator and
denominator by the least common denominator (Method 2).**

*Simplify each complex fraction by multiplying numerator and denominator by the least
common denominator.*

16. $\dfrac{\frac{9}{20}}{-\frac{11}{25}}$

16. _____

17. $\dfrac{\frac{1}{2} + \frac{3}{8}}{\frac{3}{4} - \frac{9}{8}}$

17. _____

18. $\dfrac{\frac{9}{x^2} - 1}{\frac{3}{x} - 1}$

18. _____

19. $\dfrac{\frac{16r^2}{11s^3}}{\frac{8r^4}{22s}}$

19. _____

20. $\dfrac{2x - y^2}{x + \frac{y^2}{x}}$

20. _____

**21.** $\dfrac{r + \dfrac{3}{r}}{\dfrac{5}{r} + rt}$

**22.** $\dfrac{\dfrac{x-2}{x+2}}{\dfrac{x}{x-2}}$

**23.** $\dfrac{2s + \dfrac{3}{s}}{\dfrac{1}{s} - 3s}$

**24.** $\dfrac{\dfrac{15}{10k+10}}{\dfrac{5}{3k+3}}$

**25.** $\dfrac{\dfrac{1}{h} - 4}{\dfrac{1}{2} + 2h}$

**26.** $\dfrac{\dfrac{4}{x}-\dfrac{1}{2}}{\dfrac{5}{x}+\dfrac{1}{3}}$

**26.** _____

**27.** $\dfrac{\dfrac{1}{m-1}+4}{\dfrac{2}{m-1}-4}$

**27.** _____

**28.** $\dfrac{\dfrac{4}{x+4}}{\dfrac{3}{x^2-16}}$

**28.** _____

**29.** $\dfrac{\dfrac{6}{k+1}-\dfrac{5}{k-3}}{\dfrac{3}{k-3}+\dfrac{2}{k+2}}$

**29.** _____

**30.** $\dfrac{\dfrac{4}{s+3}-\dfrac{2}{s-3}}{\dfrac{5}{s^2-9}}$

**30.** _____

# Chapter 14 RATIONAL EXPRESSIONS AND APPLICATIONS

### 14.6   Solving Equations with Rational Expressions

| Learning Objectives |
|---|
| 1    Distinguish between operations with rational expressions and equations with terms that are rational expressions. |
| 2    Solve equations with rational expressions. |
| 3    Solve a formula for a specified variable. |

### Key Terms

Use the vocabulary terms listed below to complete each statement in exercises 1–2.

**equation**          **expression**

1.   The sum or difference of terms with rational coefficients lead to an

_____.

2.   An _____ can be solved.

### Objective 1   Distinguish between operations with rational expressions and equations with terms that are rational expressions.

*Identify each of the following as an operation or an equation. Then simplify the expression or solve the equation.*

1.   $\dfrac{2x}{3} + \dfrac{2x}{5} = \dfrac{64}{15}$

1. _____

2.   $\dfrac{3x}{5} - \dfrac{4x}{3} = \dfrac{22}{15}$

2. _____

3.   $\dfrac{4x}{5} - \dfrac{5x}{10}$

3. _____

**4.**  $\dfrac{2x}{4} - \dfrac{3x}{2}$

**5.**  $\dfrac{2x}{5} + \dfrac{7x}{3}$

**6.**  $\dfrac{9}{8}p - \dfrac{1}{2} = \dfrac{1}{8}p$

**Objective 2    Solve equations with rational expressions.**

*Solve each equation and check your solutions.*

**7.**  $\dfrac{x-4}{5} = \dfrac{x+2}{3}$

**8.**  $\dfrac{4}{n+2} - \dfrac{2}{n} = \dfrac{1}{6}$

**9.**  $\dfrac{x}{3x+16} = \dfrac{4}{x}$

**10.**  $\dfrac{3}{k} - \dfrac{2}{k+2} = \dfrac{7}{3}$

**10.** _____

**11.**  $\dfrac{z+4}{z-7} - 6 = \dfrac{5}{z-7}$

**11.** _____

**12.**  $\dfrac{8}{2m+4} + \dfrac{2}{3m+6} = \dfrac{7}{9}$

**12.** _____

**13.**  $\dfrac{2}{z-1} + \dfrac{3}{z+1} - \dfrac{17}{24} = 0$

**13.** _____

**14.**  $\dfrac{5a+1}{2a+2} = \dfrac{5a-5}{5a+5} + \dfrac{3a+1}{a+1}$

**14.** _____

**15.** $\dfrac{-16}{n^2 - 8n + 12} = \dfrac{3}{n-2} + \dfrac{n}{n-6}$

**15.** _____

**16.** $\dfrac{9}{x^2 - x - 12} = \dfrac{3}{x-4} - \dfrac{x}{x+3}$

**16.** _____

**17.** $\dfrac{4}{y+2} - \dfrac{3}{y+3} = \dfrac{8}{y^2 + 5y + 6}$

**17.** _____

**18.** $\dfrac{-13}{r^2 + 6r + 8} + \dfrac{4}{r+2} = \dfrac{3}{r+4}$

**18.** _____

**19.** $\dfrac{1}{q^2 + 5q + 6} + \dfrac{1}{q^2 - 2q - 8} = \dfrac{-1}{12q + 24}$

**19.** _____

**20.**   $\dfrac{3p}{p^2+5p+6}=\dfrac{5p}{p^2+2p-3}-\dfrac{2}{p^2+p-2}$

**20.** _____

**21.**   $-\dfrac{17}{z^2+5z-6}-\dfrac{3}{1-z}=\dfrac{z+2}{z+6}$

**21.** _____

**22.**   $\dfrac{1}{x^2+5x+6}=\dfrac{2}{x^2-x-6}-\dfrac{3}{9-x^2}$

**22.** _____

**Objective 3   Solve a formula for a specified variable.**

*Solve each formula for the specified variable.*

**23.**   $S=\dfrac{a_1}{1-r}$  for $r$

**23.** _____

**24.**   $\dfrac{1}{f}=\dfrac{1}{d_0}+\dfrac{1}{d_1}$  for $f$

**24.** _____

**25.**     $m = \dfrac{y_2 - y_1}{x_2 - x_1}$ for $y_1$

    **25.** _____

**26.**     $A = \dfrac{2pf}{b(q+1)}$ for $q$

    **26.** _____

**27.**     $F = \dfrac{Gm_1m_2}{d^2}$ for $G$

    **27.** _____

**28.**     $S_n = \dfrac{n}{2}(a_1 + a_n)$ for $a_1$

    **28.** _____

**29.**     $A = \dfrac{1}{2}h(b_1 + b_2)$ for $b_2$

    **29.** _____

**30.**     $A = \dfrac{R_1R_2}{R_1 + R_r}$ for $R_r$

    **30.** _____

# Chapter 14 RATIONAL EXPRESSIONS AND APPLICATIONS

### 14.7    Applications of Rational Expressions

| **Learning Objectives** |
| --- |
| 1    Solve problems about numbers. |
| 2    Solve problems about distance, rate, and time. |
| 3    Solve problems about work. |

### Key Terms

Use the vocabulary terms listed below to complete each statement in exercises 1–3.

        **reciprocal**           **numerator**           **denominator**

1.    In the fraction $\dfrac{x+5}{x-2}$, $x+5$ is the _____.

2.    In the fraction $\dfrac{x+5}{x-2}$, $x-2$ is the _____.

3.    The fraction $\dfrac{x+5}{x-2}$ is the _____ of the fraction $\dfrac{x-2}{x+5}$.

### Objective 1    Solve problems about numbers.

*Solve each problem. Check your answers to be sure they are reasonable.*

1.    If three times a number is subtracted from twice its reciprocal, the result is –1. Find the number.

    1.    _____

2.    If twice the reciprocal of a number is added to the number, the result is $\dfrac{9}{2}$. Find the number.

    2.    _____

**3.** In a certain fraction, the numerator is 4 less than the denominator. If 5 is added to both the numerator and the denominator, the resulting fraction is equal to $\frac{7}{9}$. Find the original fraction.

**3.** _____

**4.** If a certain number is added to the numerator and twice that number is subtracted from the denominator of the fraction $\frac{3}{5}$, the result is equal to 5. Find the number.

**4.** _____

**5.** If two times a number is added to one-half of its reciprocal, the result is $\frac{13}{6}$. Find the number.

**5.** _____

**6.** If the same number is added to the numerator and denominator of the fraction $\frac{5}{9}$, the value of the resulting fraction is $\frac{2}{3}$. Find the number.

**6.** _____

**7.** The denominator of a fraction is 1 less than twice the numerator. If the numerator and the denominator are each increased by 3, the resulting fraction simplifies to $\frac{3}{4}$. Find the original fraction.

7. _____

**8.** Find two consecutive integers such that the sum of their reciprocals is $\frac{11}{30}$.

8. _____

**9.** The total resistance $R$ in an electrical circuit consisting of two resistors of $a$ ohms and $b$ ohms is given by the equation $\frac{1}{R} = \frac{1}{a} + \frac{1}{b}$. The larger resistor is 2 ohms more than the smaller resistance. Find the value of each resistor if the total resistance is two-thirds of the smaller resistance.

9. _____

**10.** In a certain fraction, the denominator is 4 more than the numerator. If 7 is added to both the numerator and the denominator, the value of the resulting fraction is $\frac{3}{4}$. Find the original fraction.

**10.** _____

**Objective 2   Solve problems about distance, rate, and time.**

*Solve each problem.*

**11.** A boat travels 15 miles per hour in still water. The boat travels 20 miles downstream in the same time it takes the boat to travel 10 miles upstream. How fast is the current?

**11.** _____

**12.** Carl traveled to his destination at an average speed of 70 miles per hour. Coming home, his average speed was 50 miles per hour and the trip took 2 hours longer. How far did he travel each way?

**12.** _____

**13.**   Dipti flew her plane 600 miles against the wind in          **13.** _____
the same time it took her to fly 900 miles with the
wind. If the speed of the wind was 30 miles per
hour, what was the speed of the plane?

**14.**   Wendy drove 250 miles to her aunt's house at a            **14.** _____
speed that was 10 miles per hour faster than her
speed on her return trip. If it took Wendy $\frac{5}{6}$ hour
longer on the return trip, what was her speed on the
return trip?

**15.**   A boat goes 6 miles per hour in still water. It takes       **15.** _____
as long to go 40 miles upstream as 80 miles
downstream. Find the speed of the current.

**16.**   A ship goes 120 miles downriver in $2\frac{2}{3}$ hours less   **16.** _____
than it takes to go the same distance upriver. If the
speed of the current is 6 miles per hour, find the
speed of the ship.

**17.** A plane traveling 450 miles per hour can go 1000 miles with the wind in $\frac{1}{2}$ hour less than when traveling against the wind. Find the speed of the wind.

**17.** _____

**18.** On Saturday, Pablo jogged 6 miles. On Monday, jogging at the same speed, it took him 30 minutes longer to cover 10 miles. How fast did Pablo jog?

**18.** _____

**19.** A plane made the trip from Redding to Los Angeles, a distance of 560 miles, in 1.5 hours less than it took to fly from Los Angeles to Portland, a distance of 1130 miles. Find the rate of the plane. (Assume there is no wind in either direction.)

**19.** _____

**20.** The Cuyahoga River has a current of 2 miles per hour. Ali can paddle 10 miles downstream in the time it takes her to paddle 2 miles upstream. How fast can Ali paddle?

**20.** _____

## Objective 3   Solve problems about work.

*Solve each problem.*

**21.** Kelly can clean the house in 6 hours, but it takes Linda 4 hours. How long would it take them to clean the house if they worked together?

21. _____

**22.** One pipe can fill a swimming pool in 8 hours and another pipe can fill the pool in 12 hours. How long will it take to fill the pool if both pipes are open?

22. _____

**23.** Chuck can weed the garden in $\frac{1}{2}$ hour, but David takes 2 hours. How long does it take them to weed the garden if they work together?

23. _____

**24.** Jack can paint a certain room in $1\frac{1}{2}$ hours, but Joe needs 4 hours to paint the same room. How long does it take them to paint the room if they work together?

**24.** _____

**25.** Michael can type twice as fast as Sharon. Together they can type a certain job in 2 hours. How long would it take Michael to type the entire job by himself?

**25.** _____

**26.** Working together, Ethel and Al can balance the books for a certain company in 3 hours. Working alone, it would take Ethel $\frac{2}{3}$ as long as Al to balance the books. How long would it take Al to do the job alone?

**26.** _____

**27.**   Judy and Tony can mow the lawn together in 4 hours. It takes Tony twice as long as Judy to do the job alone. How long would it take Judy working alone?

**27.** _____

**28.**   Fred can seal an asphalt driveway in $\frac{1}{3}$ the time it takes John. Working together, it takes them $1\frac{1}{2}$ hours. How long would it have taken Fred working alone?

**28.** _____

**29.**   A swimming pool can be filled by an inlet pipe in 18 hours and emptied by an outlet pipe in 24 hours. How long will it take to fill the empty pool if the outlet pipe is accidentally left open at the same time as the inlet pipe is opened?

**29.** _____

**30.** A cold water faucet can fill a sink in 12 minutes, and a hot water faucet can fill it in 15 minutes. The drain can empty the sink in 25 minutes. If both faucets are on and the drain is open, how long would it take to fill the sink?

**30.** _____

# Chapter 14 RATIONAL EXPRESSIONS AND APPLICATIONS

## 14.8    Variation

| Learning Objectives |
| --- |
| 1    Solve direct variation problems. |
| 2    Solve inverse variation problems. |

### Key Terms

Use the vocabulary terms listed below to complete each statement in exercises 1–3.

**direct variation        constant of variation        inverse variation**

1.    In the equation $y = kx$, the number $k$ is called the _____.

2.    If two positive quantities $x$ and $y$ are in _____
       and the constant of variation is positive, then as $x$ increases, $y$ also increases.

3.    If two positive quantities $x$ and $y$ are in _____
       and the constant of variation is positive, then as $x$ increases, $y$ decreases.

### Objective 1    Solve direct variation problems.

*Solve each problem involving direct variation.*

1.    If $m$ varies directly as $p$, and $m = 40$ when $p = 5$,          1. _____
       find $m$ when $p$ is 9.

2.    If $y$ varies directly as $x$, and $x = 14$ when $y = 42$, find          2. _____
       $y$ when $x = 4$.

3.    If $a$ varies directly as $b$, and $a = 24$ when $b = 16$, find          3. _____
       $b$ when $a = 34$.

4.    If $c$ varies directly as $d$, and $c = 100$ when $d = 5$, find          4. _____
       $c$ when $d = 3$.

**5.** If $h$ varies directly as $m$, and $h = 9$ when $m = 6$, find        **5.** _____
    $h$ when $m = 10$.

**6.** If $x$ varies directly as $y$, and $x = 24$ when $y = -2$,           **6.** _____
    find $x$ when $y = 4$.

**7.** If $f$ varies directly as $g$, and $f = \frac{7}{2}$ when $g = 7$, find $f$        **7.** _____
    when $g = 18$.

**8.** If $w$ varies directly as $v$, and $w = 24$ when $v = 20$,           **8.** _____
    find $w$ when $v = 25$.

**9.** If $a$ varies directly as $b$, and $a = 61.5$ when $b = 82$,         **9.** _____
    find $a$ when $b = 224$.

**10.** If $y$ varies directly as $x$, and $y = 21$ when $x = 35$, find     **10.** _____
    $y$ when $x = 75$.

## Objective 2    Solve inverse variation problems.

*Solve each problem involving indirect variation.*

**11.** If $y$ varies inversely as $x$, and $y = 20$ when $x = 4$,          **11.** _____
    find $y$ when $x = 12$.

**12.** If $y$ varies inversely as $x$, and $y = 5$ when $x = 3$, find $y$ when $x = 0.5$.

**12.** _____

**13.** If $g$ varies inversely as $f$, and $g = 6$ when $f = 12$, find $g$ when $f = 18$.

**13.** _____

**14.** If $d$ varies inversely as $c$, and $d = 18$ when $c = \dfrac{1}{3}$, find $d$ when $c = \dfrac{2}{5}$.

**14.** _____

**15.** If $n$ varies inversely as $m$, and $n = 10.5$ when $m = 1.2$, find $n$ when $m = 5.6$.

**15.** _____

**16.** If $y$ varies inversely as $x$, and $y = 10$ when $x = 2$, find $y$ when $x = 4$.

**16.** _____

**17.** If $a$ varies inversely as $b$, and $a = 36$ when $b = \dfrac{1}{2}$, find $a$ when $b = 3$.

**17.** _____

**18.**  If $y$ varies inversely as $x$, and $y = 9$ when $x = 2$, find  **18.** _____
$y$ when $x = 4$.

**19.**  If $c$ varies inversely as $d$, and $c = 11$ when $d = 3$, find  **19.** _____
$c$ when $d = 11$.

**20.**  If $t$ varies inversely as $s$, and $t = 100$ when $s = 0.4$,  **20.** _____
find $t$ when $s = 6$.

*Solve each problem*

**21.**  For a given period of time, the interest earned on an  **21.** _____
investment varies directly as the interest rate. If the
interest is $125 when the rate is 5%, find the interest
when the rate is $6\frac{1}{2}\%$.

**22.**  For a specified distance, time varies inversely with  **22.** _____
speed. If Ramona walks a certain distance on a
treadmill in 40 minutes at 4.2 miles per hour, how
long will it take her to walk the same distance at 3.5
miles per hour?

**23.** If the temperature is constant, the pressure of a gas in a container varies inversely as the volume of the container. If the pressure is 140 pounds per square foot when a gas is in a container of 1000 cubic feet, what is the volume of the container when the gas exerts a pressure of 700 pounds per square feet?

**23.** _____

**24.** If the temperature is constant, the pressure of a gas in a container varies inversely as the volume of the container. If the pressure is 9 pounds per square foot in a container of 6 cubic feet, what is the pressure in a container of 7.5 cubic feet?

**24.** _____

**25.** The circumference of a circle varies directly as the radius. A circle with a radius of 7 centimeters has a circumference of 43.96 centimeters. Find the circumference if the radius changes to 11 centimeters.

**25.** _____

**26.** For a constant area, the length of a rectangle varies inversely as the width. If the length of a rectangle is 16 feet when the width is 3 feet, find the length of a rectangle with the same area where the width is 9 feet.

**26.** _____

**27.** For a given height, the area of a triangle varies directly as its base. Find the area of a triangle with a base of 4 centimeters, if the area is 9.6 square centimeters when the base is 3 centimeters.

27. _____

**28.** The force required to compress a spring varies directly as the change in the length of the spring. If a force of 25 pounds is required to compress a spring 4 inches, how much force is required to compress the spring 8 inches?

28. _____

**29.** For a given rate, the distance that an object travels varies directly with time. Find the distance an object travels in 5 hours if the object travels 165 miles in 3 hours.

29. _____

**30.** The length of a violin string varies inversely with the frequency of its vibrations. A 10-inch violin string vibrates at a frequency of 512 cycles per second. Find the frequency of an 8-inch string.

30. _____

# Chapter 15 SYSTEMS OF LINEAR EQUATIONS AND INEQUALITIES

## 15.1    Solving Systems of Linear Equations by Graphing

| Learning Objectives |
| --- |
| 1    Decide whether a given ordered pair is a solution of a system. |
| 2    Solve linear systems by graphing. |
| 3    Solve special systems by graphing. |
| 4    Identify special systems without graphing. |

## Key Terms

Use the vocabulary terms listed below to complete each statement in exercises 1–7.

| | |
| --- | --- |
| **system of linear equations** | **solution of the system** |
| **solution set of the system** | **consistent system**    **inconsistent system** |
| **independent equations** | **dependent equations** |

1.    Equations of a system that have different graphs are called
      _____.

2.    A system of equations with at least one solution is a
      _____.

3.    The set of all ordered pairs that are solutions of a system is the
      _____.

4.    The _____ of linear equations includes all
      the ordered pairs that make all the equations of the system true at the same time.

5.    Equations of a system that have the same graph (because they are different forms
      of the same equation) are called _____.

6.    A system with no solution is called a(n) _____.

7.    A(n) _____ consists of two or more linear
      equations with the same variables.

**Objective 1   Decide whether a given ordered pair is a solution of a system.**

*Decide whether the given ordered pair is a solution of the given system.*

1.   $(4,1)$

$2x+3y=11$

$3x-2y=9$

1. _____

2.   $(2,-4)$

$2x+3y=6$

$3x-2y=14$

2. _____

3.   $(-3,-1)$

$5x-3y=-12$

$2x+3y=-9$

3. _____

4.   $(4,0)$

$4x+3y=16$

$x-4y=-4$

4. _____

5.   $(-5,-4)$

$x-y=-1$

$4x+y=-24$

5. _____

6.   $(3,-7)$

$5x+\ y=8$

$2x-3y=26$

6. _____

7.   $(-1,-7)$

$x-\ y=6$

$-2x+3y=-19$

7. _____

## Objective 2   Solve linear systems by graphing.

*Solve each system by graphing both equations on the same axes.*

**8.**   $x - 2y = 6$

  $2x + y = 2$

**8.**

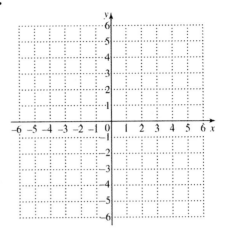

_____

**9.**   $2x + 3y = 5$

  $3x - y = 13$

**9.**

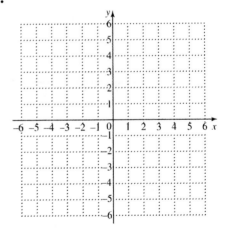

_____

**10.** $6x - 5y = 4$

$\quad\quad 2x - 5y = 8$

**10.**

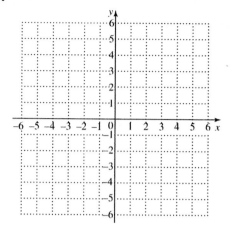

**11.** $3x - y = -7$

$\quad\quad 2x + y = -3$

**11.**

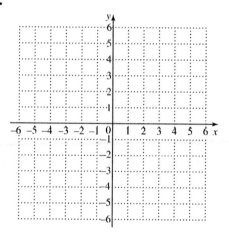

**12.** $\quad\quad 2x = y$

$\quad\ 5x + 3y = 0$

**12.**

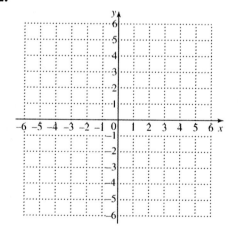

**13.**     $y - 2 = 0$

           $3x - 4y = -17$

**13.**

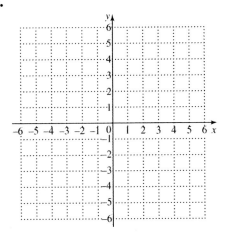

**14.**     $3x + 2 = y$

           $2x - y = 0$

**14.**

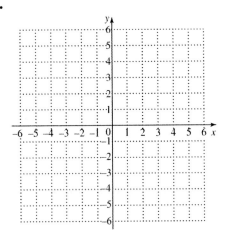

**15.**     $x - y = -7$

           $x + 11 = 2y$

**15.**

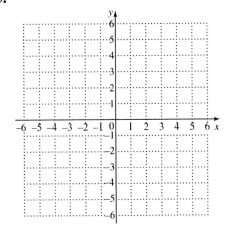

## Objective 3    Solve special systems by graphing.

*Solve each system of equations by graphing both equations on the same axes. If the two equations produce parallel lines, write **no solution**. If the two equations produce the same line, write **infinite number of solutions**.*

**16.**    $8x + 4y = -1$

$4x + 2y = 3$

**16.**

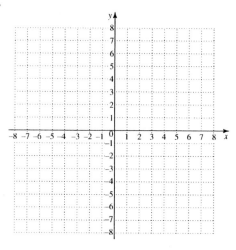

_____

**17.**    $x + 2y = 4$

$8y = -4x + 16$

**17.**

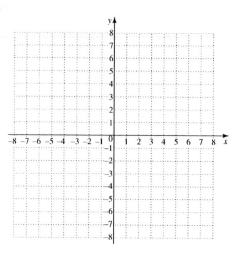

_____

**18.**     $4x + 3y = 12$

        $6y + 8x = -24$

**18.**

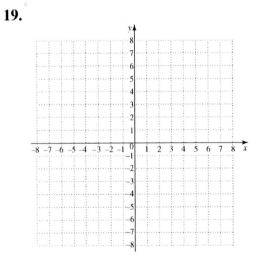

_____

**19.**     $2x + 3y = 0$

        $6x = -9y$

**19.**

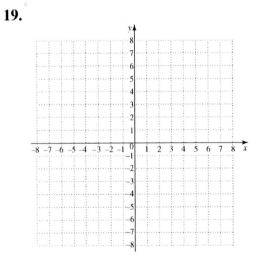

_____

**20.**     $-3x + 2y = 6$

        $-6x + 4y = 12$

**20.**

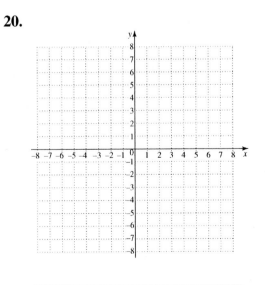

_____

**21.**     $3x + 3y = 8$                 **21.**

$x = 4 - y$

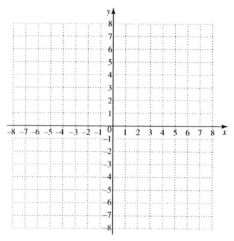

_____

**22.**     $5x + 3y = 30$               **22.**

$10x + 6y = 60$

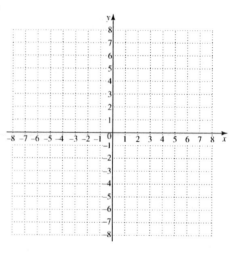

_____

**Objective 4    Identify special systems without graphing.**

*Without graphing, answer the following equations for each linear system.*
*(a) Is the system inconsistent, are the equations dependent, or neither?*
*(b) Is the graph a pair of intersecting lines, a pair of parallel lines, or one line?*
*(c) Does the system have one solution, no solution, or an infinite number of solutions?*

**23.**       $y = 2x + 1$               **23.  (a)**_____

$3x - y = 7$

**(b)** _____

**(c)** _____

**24.** $-2x + \phantom{2}y = 4$

$\phantom{-}-4x + 2y = -2$

**24.** (a)_____

(b)_____

(c)_____

**25.** $4x + 3y = 12$

$\phantom{4x}-12x = -36 + 9y$

**25.** (a)_____

(b)_____

(c)_____

**26.** $x - y = 1$

$\phantom{x-}y = x - 1$

**26.** (a)_____

(b)_____

(c)_____

**27.** $\phantom{3x-2}y = -2x + 5$

$3x - 2y = 4$

**27.** (a)_____

(b)_____

(c)_____

**28.** $y = \dfrac{1}{2}x - 4$

$2y = x - 8$

**28.** (a)_____

(b)_____

(c)_____

**29.**   $3x + 5y = -5$

   $2y + 5 = 0$

**29. (a)** _____

   **(b)** _____

   **(c)** _____

**30.**   $y = \dfrac{1}{3}x + 2$

   $3y = x - 21$

**30. (a)** _____

   **(b)** _____

   **(c)** _____

# Chapter 15 SYSTEMS OF LINEAR EQUATIONS AND INEQUALITIES

## 15.2   Solving Systems of Linear Equations by Substitution

| **Learning Objectives** |
| --- |
| 1   Solve linear systems by substitution. |
| 2   Solve special systems by substitution. |
| 3   Solve linear systems with fractions and decimals by substitution. |

## Key Terms

Use the vocabulary terms listed below to complete each statement in exercises 1–4.

**substitution**          **ordered pair**          **inconsistent system**

**dependent system**

1.   The solution of a linear system of equations is written as a(n)
     _____.

2.   When one expression is replaced by another, _____ is being
     used.

3.   A system of equations in which all solutions of the first equation are also
     solutions of the second equation is a(n) _____.

4.   A system of equations that has no common solution is called a(n)
     _____.

## Objective 1   Solve linear systems by substitution.

*Solve each system by the substitution method. Check each solution.*

1.   $x + y = 7$
     $y = 6x$                                    1. _____

2.   $3x + 2y = 14$
     $y = x + 2$                                 2. _____

**3.**    $x + y = 9$

$5x - 2y = -4$

**3.** _____

**4.**    $x - 4y = 17$

$3x - 4y = 11$

**4.** _____

**5.**    $-8x + 5y = 11$

$x - y = -1$

**5.** _____

**6.**    $5x + y = 8$

$10x - 3y = -19$

**6.** _____

**7.**    $x - y = 6$

$-2x + 3y = -19$

**7.** _____

**8.**    $3x - 21 = y$

$y + 2x = -1$

**8.** _____

**9.**    $x + 6y = -1$

$-2x - 9y = 0$

**9.** _____

**10.**     $2x + 3y = -4$

$3x - 2y = 7$

**10.** _____

**11.**     $2x - 5y = 11$

$3x - 4y = 6$

**11.** _____

**12.**     $2x + 4y = -1$

$-4x - 6y = 1$

**12.** _____

**Objective 2     Solve special systems by substitution.**

*Solve each system by the substitution method. Use set-builder notation for dependent equations.*

**13.**     $y = -\dfrac{1}{3}x + 5$

$3y + x = -9$

**13.** _____

**14.**     $\dfrac{1}{2}x + 3 = y$

$6 = -x + 2y$

**14.** _____

**15.**     $3x + 5y = 7$

$6x + 10y = 3$

**15.** _____

**16.**　　$4x + 3y = 2$

　　　　$8x + 6y = 4$

**17.**　　$3x + 5y = 3$

　　　　$6x + 10y = -2$

17.　_____

**18.**　　$4x - 2y = 3$

　　　　$-8x + 4y = -6$

18.　_____

**Objective 3　Solve linear systems with fractions and decimals by substitution.**

*Solve each system by the substitution method. Check each solution*

**19.**　　$\dfrac{5}{3}x + y = 12$

　　　　$x + \dfrac{1}{2}y = 7$

19.　_____

**20.**　　$\dfrac{5}{4}x - y = -\dfrac{1}{4}$

　　　　$-\dfrac{7}{8}x + \dfrac{5}{8}y = 1$

20.　_____

**21.** $x + \dfrac{5}{3} y = 7$

$\dfrac{5}{6} x + \dfrac{2}{3} y = \dfrac{11}{3}$

**21.** _____

**22.** $x - \dfrac{3}{4} y = 3$

$\dfrac{1}{4} x - \dfrac{1}{2} y = \dfrac{1}{8}$

**22.** _____

**23.** $\dfrac{1}{20} x - \dfrac{1}{15} y = \dfrac{1}{6}$

$\dfrac{1}{6} x + y = 3$

**23.** _____

**24.** $\dfrac{1}{4} x + \dfrac{3}{8} y = -3$

$\dfrac{5}{6} x - \dfrac{3}{7} y = -10$

**24.** _____

**25.**  $0.1x + 0.3y = 0.1$

$0.2x = -1 - 1.2y$

**25.** _____

**26.**  $0.2y + 0.8x = -0.2$

$0.1x + 0.3y = -1.4$

**26.** _____

**27.**  $0.3x - 2.1 = 0.1y$

$0.3y + 0.6x = -0.3$

**27.** _____

**28.**  $0.3x + 1.8y = -2.7$

$-0.2x - 0.6y = 0$

**28.** _____

**29.**  $0.6x + 0.8y = 1$

$0.4y = 0.5 - 0.3x$

**29.** _____

**30.**  $0.4x + 0.3y = -1$

$0.2x + 0.3y = -0.2$

**30.** _____

# Chapter 15 SYSTEMS OF LINEAR EQUATIONS AND INEQUALITIES

### 15.3   Solving Systems of Linear Equations by Elimination

| Learning Objectives |
| --- |
| 1    Solve linear systems by elimination. |
| 2    Multiply when using the elimination method. |
| 3    Use an alternative method to find the second value in a solution. |
| 4    Use the elimination method to solve special systems. |

### Key Terms

Use the vocabulary terms listed below to complete each statement in exercises 1–3.

**addition property of equality**      **elimination method**      **substitution**

1.   Using the addition property to solve a system of equation is called the

    _____.

2.   The _____ states that the same quantity to each side
    of an equation results in equal sums.

3.   _____ is being used when one expression is replaced by
    another,.

### Objective 1   Solve linear systems by elimination.

*Solve each system by the elimination method. Check your answers.*

1.   $x + y = 5$
     $x - y = -3$

1. _____

2.   $3x - y = 5$
     $2x + y = 0$

2. _____

**3.**     $x - 4y = -4$

   $-x + y = -5$

**4.**     $2x - y = 10$

   $3x + y = 10$

**5.**     $4x + 3y = -4$

   $2x - 3y = 16$

**6.**     $8x + 2y = 14$

   $3x - 2y = -14$

**7.**     $x - 3y = 5$

   $-x + 4y = -5$

**Objective 2     Multiply when using the elimination method.**

*Solve each system by the elimination method. Check your answers.*

**8.**     $6x + 7y = 10$

   $2x - 3y = 14$

**9.**  $8x + 6y = 10$

  $4x - \;y = 1$

**9.** _____

**10.**  $3x + 2y = 5$

  $2x - 3y = 12$

**10.** _____

**11.**  $3x + 5y = 8$

  $2x - y = -12$

**11.** _____

**12.**  $6x + y = 1$

  $3x - 4y = 23$

**12.** _____

**13.**  $4x - 5y = -22$

  $3x + 2y = -5$

**13.** _____

**14.**  $4x - 9y = 7$

  $3x + 2y = 14$

**14.** _____

**15.** $3x - 7y = 12$

$5x + 3y = -2$

**15.** _____

**Objective 3    Use an alternative method to find the second value in a solution.**

*Solve each system by the elimination method. Check your answers.*

**16.** $5x - 3y = 23$

$10 + 2y = 2x$

**16.** _____

**17.** $4y = 2x - 2$

$-9 + 3y = 5x$

**17.** _____

**18.** $4x - 3y - 20 = 0$

$6x + 5y + 8 = 0$

**18.** _____

**19.** $5x + 5y + 15 = 0$

$3x + 4y = -8$

**19.** _____

**20.** $6x = 16 - 7y$

$4x = 3y + 26$

**20.** _____

**21.** $2x = 14 + 4y$

$6y = -5x + 3$

**21.** _____

**22.** $7 - y = 2x$

$4x = 19 + 3y$

**22.** _____

**23.** $5x + 3y + 4 = 0$

$4x + 5y - 2 = 0$

**23.** _____

**24.** $2x = 21 - 3y$

$\frac{1}{3}x + \frac{2}{5}y = 3$

**24.** _____

**Objective 4    Use the elimination method to solve special systems.**

*Solve each system by the elimination method. Use set-builder notation for dependent equations. Check your answers.*

**25.** $12x - 8y = 3$

$6x - 4y = 6$

**25.** _____

**26.**  $2x + 4y = -6$

$-x - 2y = 3$

**26.** _____

**27.**  $6x - 12y = 3$

$2x - 4y = 1$

**27.** _____

**28.**  $15x + 6y = 9$

$10x + 4y = 18$

**28.** _____

**29.**  $48x - 56y = 32$

$21y - 18x = -12$

**29.** _____

**30.**  $15x - 10y = 6$

$-12x + 8y = 2$

**30.** _____

# Chapter 15 SYSTEMS OF LINEAR EQUATIONS AND INEQUALITIES

## 15.4    Applications of Linear Systems

| **Learning Objectives** |
| :--- |
| 1       Solve problems about unknown numbers. |
| 2       Solve problems about quantities and their costs. |
| 3       Solve problems about mixtures. |
| 4       Solve problems about distance, rate (or speed), and time. |

## Key Terms

Use the vocabulary terms listed below to complete each statement in exercises 1–2.

**system of linear equations**          $d = rt$

1.    The formula that relates distance, rate, and time is _____.

2.    A _____ consists of at least two linear equations
      with different variables.

## Objective 1    Solve problems about unknown numbers.

*Write a system of equations for each problem, then solve the problem.*

1.    The sum of two numbers is 20. Three times the
      smaller is equal to twice the larger. Find the
      numbers.

      **1.**

      larger number _____

      smaller number _____

2.    The difference between two numbers is 14. If two
      times the smaller is added to one-half the larger, the
      result is 52. Find the numbers.

      **2.**

      larger number _____

      smaller number _____

3.    Two towns have a combined population of 9045.
      There are 2249 more people living in one than in the
      other. Find the population in each town.

      **3.**

      larger town _____

      smaller town _____

**4.** There are a total of 49 students in the two second grade classes at Jefferson School. If Carla has 7 more students in her class than Linda, find the number of students in each class.

**4.**

Carla's class _____

Linda's class _____

**5.** A rope 82 centimeters long is cut into two pieces with one piece four more than twice as long as the other. Find the length of each piece.

**5.**

longer piece _____

shorter piece _____

**6.** The perimeter of a rectangular room is 50 feet. The length is three feet greater than the width. Find the dimensions of the rectangle.

**6.**

length _____

width _____

**7.** The perimeter of a triangular pennant is 116 centimeters. If two sides are of equal length, and the third side is 20 centimeters longer than each of the equal sides, what are the lengths of the three sides?

**7.**

side 1_____

side 2_____

side 3_____

## Objective 2   Solve problems about quantities and their costs.

*Write a system of equations for each problem, then solve the problem.*

**8.** There were 411 tickets sold for a soccer game, some for students and some for nonstudents. Student tickets cost $4.25 and nonstudent tickets cost $8.50 each. The total receipts were $3021.75. How many of each type were sold?

**8.**

student tix _____

nonstudent tix _____

**9.** A cashier has some $5 bills and some $10 bills. The total value of the money is $750. If the number of tens is equal to twice the number of fives, how many of each type are there?

**9.**

$5 bills _____

$10 bills _____

**10.** The total receipts for a basketball game were $4690.50. There were 723 tickets sold, some for children and some for adults. If the adult tickets cost $9.50 and the children's tickets cost $4, how many of each type were there?

**10.**

adult tix_____

children tix _____

**11.** Wendy has $10,000 to invest, part at 7% and part at 4%. If the total annual income from simple interest is to be $580, how much should she invest at each rate?

**11.**

7% amount _____

4% amount _____

12.  Twice as many general admission tickets to a
basketball game were sold as reserved seat tickets.
General admission tickets cost $10 and reserved seat
tickets cost $15. If the total value of both kinds of
tickets was $26,250, how many tickets of each kind
were sold?

**12.**

general admission _____

reserved seats _____

13.  Carla has $12,000 to invest at 7% and 9%. She
wants the income from simple interest on the two
investments to total $1000 yearly. How much should
she invest at each rate?

**13.**

7% amount _____

9% amount _____

14.  Stan has 14 bills in his wallet worth $95 altogether.
If the wallet contains only $5 and $10 bills, how
many bills of each denomination does he have?

**14.**

$5 bills _____

$10 bills _____

15.  Luke plans to buy 10 ties with exactly $162. If some
ties cost $14, and the others cost $25, how many ties
of each price should he buy?

**15.**

$14 ties _____

$25 ties _____

## Objective 3    Solve problems about mixtures.

*Write a system of equations for each problem, then solve the problem.*

**16.**    Jorge wishes to make 150 pounds of coffee blend that can be sold for $8 per pound. The blend will be a mixture of coffee worth $6 per pound and coffee worth $12 per pound. How many pounds of each kind of coffee should be used in the mixture?

**16.**

$6 coffee _____

$12 coffee _____

**17.**    How many liters of water should be added to 25% antifreeze solution to get 30 liters of a 20% solution? How many liters of 25% solution are needed?

**17.**

water _____

25% solution_____

**18.**    Ben wishes to blend candy selling for $1.60 a pound with candy selling for $2.50 a pound to get a mixture that will be sold for $1.90 a pound. How many pounds of the $1.60 and the $2.50 candy should be used to get 30 pounds of the mixture?

**18.**

$1.60 candy _____

$2.50 candy _____

**19.**    How many bags of coffee worth $90 a bag must be mixed with coffee worth $75 a bag to get 50 bags worth $87 a bag?

**19.**

$90 coffee _____

$75 coffee _____

**20.** How many liters of 75% solution should be mixed with a 55% solution to get 70 liters of 63% solution? How many liters of the 55% and 75% solutions should be used?

**20.**
55% solution_____

75% solution_____

**21.** A pharmacist wants to add water to a solution that contains 80% medicine. She wants to obtain 12 oz. of a solution that is 20% medicine. How much water and how much of the 80% solution should she use?

**21.**
water_____

80% solution_____

**22.** Three quarts of a 24% iodine solution were mixed with a 52% solution to make a 40% iodine solution. How many quarts of the 52% solution were needed?

**22.**
52% solution_____

42% solution_____

**Objective 4    Solve problems about distance, rate (or speed), and time.**

*Write a system of equations for each problem, then solve the problem.*

**23.** Bill and Hillary start in Washington and fly in opposite directions. At the end of 4 hours, they are 4896 kilometers apart. If Bill flies 60 kilometers per hour faster than Hillary, what are their speeds?

**23.**
Bill  _____

Hillary_____

**24.** It takes Carla's boat $\frac{1}{2}$ hour to go 8 miles downstream and 1 hour to make the return trip upstream. Find the speed of the current and the speed of Carla's boat in still water.

**24.**
boat speed _____

current speed _____

**25.** Enid leaves Cherry Hill, driving by car toward New York, which is 90 miles away. At the same time, Jerry, riding his bicycle, leaves New York cycling toward Cherry Hill. Enid is traveling 28 miles per hour faster than Jerry. They pass each other $1\frac{1}{2}$ hours later. What are their speeds?

**25.**
Enid _____

Jerry _____

**26.** Two planes left Philadelphia traveling in opposite directions. Plane A left 15 minutes before plane B. After plane B had been flying for 1 hour, the planes were 860 miles apart. What were the speeds of the two planes if plane A was flying 40 miles per hour faster than plane B?

**26.**
plane A _____

plane B _____

**27.** John left Louisville at noon on the same day that Mike left Louisville at 1 P.M. Both were traveling in the same direction. At 5 P.M., Mike was 62 miles behind John. If John was traveling 2 miles per hour faster than Mike, what were their speeds?

**27.**
John _____

Mike _____

**28.**   It takes a kayak $1\frac{1}{2}$ hours to go 24 miles downstream and 4 hours to return. Find the speed of the current and the speed of the kayak in still water.

**28.**

kayak speed _____

current speed_____

**29.**   A plane can travel 300 miles per hour with the wind and 230 miles per hour against the wind. Find the speed of the wind and the speed of the plane in still air.

**29.**

plane speed _____

wind speed _____

**30.**   At the beginning of a fund-raising walk, Steve and Vic are 30 miles apart. If they leave at the same time and walk in the same direction, Steve would overtake Vic in 15 hours. If they walked toward each other, they would meet in 3 hours. What are their speeds?

**30.**

Steve _____

Vic   _____

# Chapter 15 SYSTEMS OF LINEAR EQUATIONS AND INEQUALITIES

### 15.5    Solving Systems of Linear Inequalities

**Learning Objectives**

1       Solve systems of linear inequalities by graphing.

### Key Terms

Use the vocabulary terms listed below to complete each statement in exercises 1–2.

**system of linear inequalities**

**solution set of a system of linear inequalities**

1.   All points that make all inequalities of the system true at the same time is called the _____.

2.   A _____ contains two or more linear inequalities (and no other kinds of inequalities).

### Objective 1    Solve systems of linear inequalities by graphing.

*Graph the solution of each system of linear inequalities.*

1.   $7x + 3y \geq 21$
     $x - y \leq 6$

1.

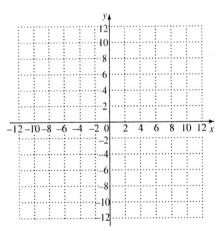

**2.**   $3x - y \leq 3$

$x + y \leq 0$

**2.**

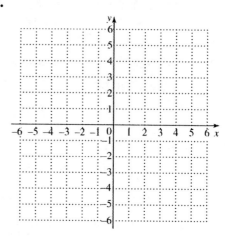

**3.**   $3x - y \leq 6$

$3y - 6 \leq 2x$

**3.**

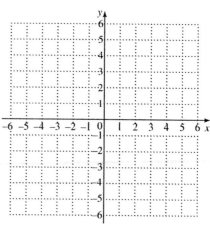

**4.**   $3x + 5y \geq 15$

$y \geq x - 2$

**4.**

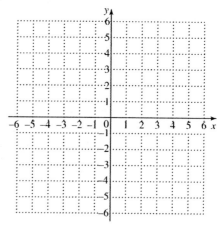

Name:

Instructor:

Date:

Section:

**5.** $x + y \leq 3$

$5x - y \geq 5$

**5.**

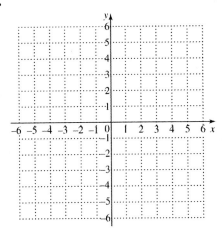

**6.** $x + 2y \geq -4$

$5x \leq 10 - 2y$

**6.**

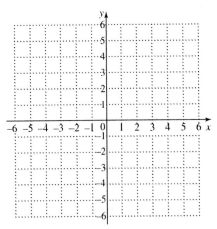

**7.** $3x - y > 3$

$4x + 3y < 12$

**7.**

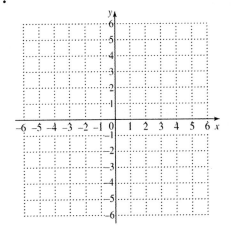

**8.**  $4x + 5y \le 20$

$y \le x + 3$

**8.**

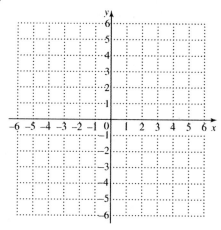

**9.**  $2x - y \ge 4$

$5y + 15 \ge -3x$

**9.**

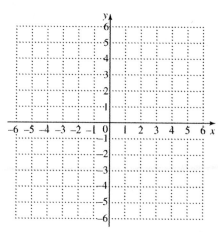

**10.**  $3x - 2y < 8$

$x < 4$

**10.**

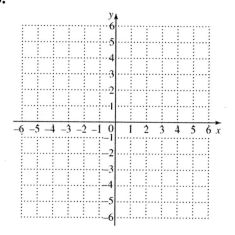

**11.**  $x - y \leq 2$

$\qquad y \leq 2$

**11.**

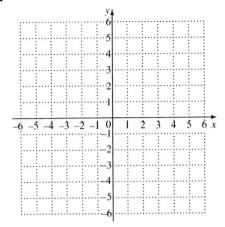

**12.**  $x + \ y \geq 3$

$\quad x - 2y \leq 4$

**12.**

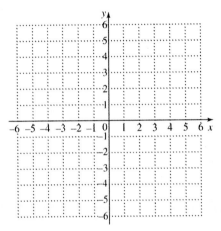

**13.**  $x < 2y + 3$

$\quad 0 < x + y$

**13.**

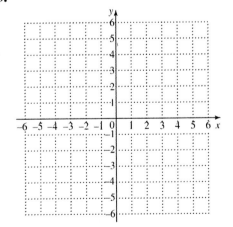

**14.**    $6x - y > 6$

        $2x + 5y < 10$

**14.**

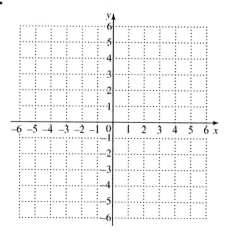

**15.**    $3x - 4y < 12$

            $y > -4$

**15.**

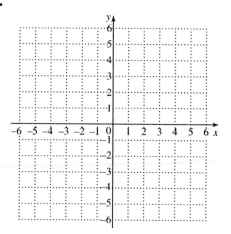

**16.**    $x - 2y \le 4$

        $x + 2y \le 4$

**16.**

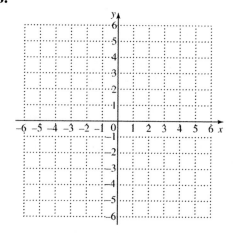

**17.**        $y > 2$

$4x - 3y < 9$

**17.**

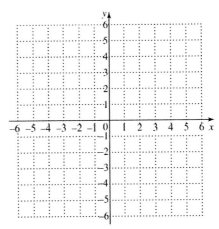

**18.**    $4x - 3y > 12$

$x < 4$

**18.**

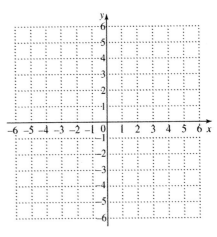

**19.**    $4x - y \leq 4$

$7y + 14 \geq -2x$

**19.**

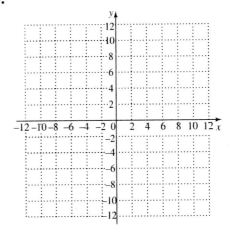

**20.**　　$5x - 2y \leq 10$

　　　　　　$y \leq -2$

**20.**

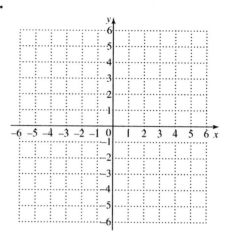

**21.**　　$x - 2y \leq 3$

　　　　　$2x + y \leq -4$

**21.**

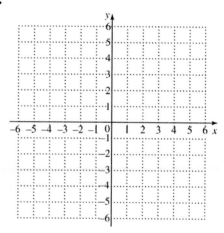

**22.**　　$x + y > -3$

　　　　　$2x - 3y \leq -2$

**22.**

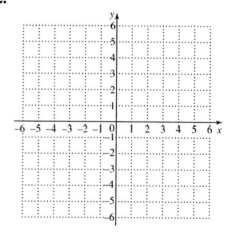

**23.**    $y < 4$

$x \geq -3$

**23.**

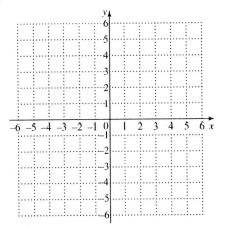

**24.**    $x - 2y \geq -7$

$x - 2y < 2$

**24.**

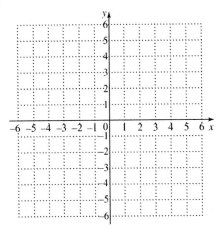

**25.**    $4x - y > 2$

$y > -x - 2$

**25.**

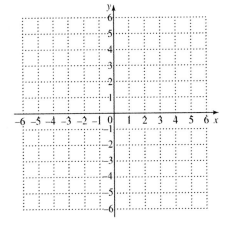

**26.**　　$3x + 2y < 10$

　　　　$5x - 2y \le 6$

**26.**

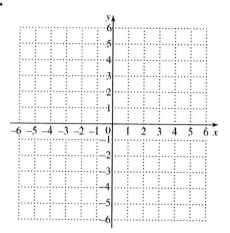

**27.**　　　　$y \ge -1$

　　　　$2x - y > -1$

**27.**

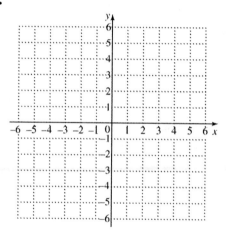

**28.**　　$x - 3y \le -7$

　　　　　$x < 2$

**28.**

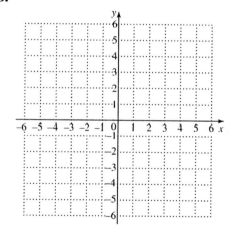

**29.**     $2x - y > -4$
$2x + y > 0$

**29.**

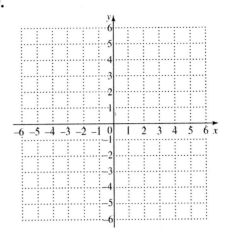

**30.**     $x - 3y \geq -9$
$x + 3y < 3$

**30.**

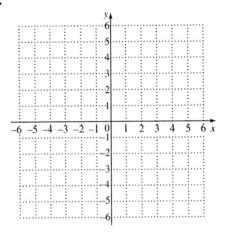

# Chapter 16 ROOTS AND RADICALS

## 16.1    Evaluating Roots

| **Learning Objectives** |
| :--- |
| 1    Find square roots. |
| 2    Decide whether a given root is rational, irrational, or not a real number. |
| 3    Find decimal approximations for irrational square roots. |
| 4    Use the Pythagorean formula. |
| 5    Find cube, fourth, and other roots. |

### Key Terms

Use the vocabulary terms listed below to complete each statement in exercises 1–9.

| square root | principal square root | radicand |
| :--- | :--- | :--- |
| radical | radical expression | perfect square |
| irrational number | cube root | index (order) |

1.    The number or expression inside a radical sign is called the _____.

2.    A number with a rational square root is called a _____.

3.    In a radical of the form $\sqrt[n]{a}$, the number $n$ is the _____.

4.    The number $b$ is a _____ of $a$ if $b^2 = a$.

5.    The expression $\sqrt[n]{a}$ is called a _____.

6.    The positive square root of a number is its _____.

7.    A real number that is not rational is called an _____.

8.    A _____ is a radical sign and the number or expression in it.

9.    The number $b$ is a _____ of $a$ if $b^3 = a$.

### Objective 1    Find square roots.

*Find all square roots of each number.*

1.    625

1.    _____

2.    $\dfrac{49}{144}$

2.    _____

**3.**  $\dfrac{121}{196}$

3. _____

*Find each square root*

**4.**  $\sqrt{400}$

4. _____

**5.**  $\sqrt{\dfrac{900}{49}}$

5. _____

**6.**  $-\sqrt{\dfrac{2500}{6400}}$

6. _____

**Objective 2**  **Decide whether a given root is rational, irrational, or not a real number.**

*Tell whether each square root is* rational, irrational, *or* not a real number.

**7.**  $\sqrt{72}$

7. _____

**8.**  $\sqrt{-36}$

8. _____

**9.**  $-\sqrt{\dfrac{625}{484}}$

9. _____

**10.**  $\sqrt{6400}$

10. _____

**11.**  $\sqrt{8.1}$

11. _____

**12.**  $\sqrt{-49}$

12. _____

**Objective 3**  **Find decimal approximations for irrational square roots.**

*Use a calculator to find a decimal approximation for each square root. Round answers to the nearest thousandth.*

**13.**  $\sqrt{32}$

13. _____

**14.** $-\sqrt{18}$

**15.** $\sqrt{200}$

**16.** $-\sqrt{131}$

**17.** $\sqrt{210}$

**18.** $-\sqrt{26,358}$

**14.** _____

**15.** _____

**16.** _____

**17.** _____

**18.** _____

**Objective 4   Use the Pythagorean formula.**

*Find the length of the unknown side of each right triangle with sides a, b, and c, where c is the hypotenuse. If necessary, round your answer to the nearest thousandth.*

**19.** $a = 5, b = 9$

**19.** _____

**20.** $c = 15, a = 12$

**20.** _____

**21.** $c = 15, b = 14$

**21.** _____

*Use the Pythagorean formula to solve each problem. If necessary, round your answer to the nearest thousandth.*

**22.** Susan started to drive due south at the same time John started to drive due west. John drove 21 miles in the same time that Susan drove 28 miles. How far apart were they at that time?

**22.** _____

**23.** A ladder 25 feet long leans against a wall. The foot of the ladder is 7 feet from the base of the wall. How high up the wall does the top of the ladder rest?

**23.** _____

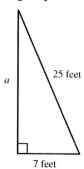

**24.** A plane flies due east for 35 miles and then due south until it is 37 miles from its starting point. How far south did the plane fly?

**24.** _____

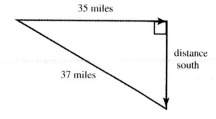

**Objective 5    Find cube, fourth, and other roots.**

*Find each root.*

**25.** $\sqrt[3]{-64}$

**25.** _____

**26.** $\sqrt[4]{-625}$

**26.** _____

**27.**  $-\sqrt[5]{243}$

**27.** _____

**28.**  $-\sqrt[3]{-343}$

**28.** _____

**29.**  $\sqrt[4]{256}$

**29.** _____

**30.**  $\sqrt[7]{-1}$

**30.** _____

# Chapter 16 ROOTS AND RADICALS

## 16.2   Multiplying, Dividing, and Simplifying Radicals

| **Learning Objectives** |
| --- |
| 1       Multiply square root radicals. |
| 2       Simplify radicals using the product rule. |
| 3       Simplify radicals using the quotient rule. |
| 4       Simplify radicals involving variables. |
| 5       Simplify other roots. |

### Key Terms

Use the vocabulary terms listed below to complete each statement in exercises 1–3.

    **perfect cube**        **radical**        **radicand**

1.    A root of a number is called a _____.

2.    A number with a rational cube root is called a _____.

3.    The _____ is the number or expression inside a radical sign.

### Objective 1   Multiply square root radicals.

*Use the product rule for radicals to find each product.*

1.    $\sqrt{13} \cdot \sqrt{5}$                         1. _____

2.    $\sqrt{7} \cdot \sqrt{7}$                          2. _____

3.    $\sqrt{10} \cdot \sqrt{30}$                     3. _____

4.    $\sqrt{13} \cdot \sqrt{8}$                       4. _____

**5.**   $\sqrt{3x} \cdot \sqrt{7},\ x > 0$

**6.**   $\sqrt{18a} \cdot \sqrt{2b},\ a \geq 0,\ b \geq 0$

**Objective 2    Simplify radicals using the product rule.**

*Simplify each radical.*

**7.**   $-\sqrt{180}$

**8.**   $\sqrt{405}$

**9.**   $3\sqrt{1000}$

*Find each product and simplify.*

**10.**   $\sqrt{11} \cdot \sqrt{33}$

**11.**   $\sqrt{20} \cdot \sqrt{10}$

**12.** $\sqrt{18} \cdot \sqrt{24}$

**12.** _____

**Objective 3    Simplify radicals using the quotient rule.**

*Use the quotient rule and product rule, as necessary to simplify each expression.*

**13.** $\sqrt{\dfrac{25}{81}}$

**13.** _____

**14.** $\sqrt{\dfrac{13}{121}}$

**14.** _____

**15.** $\dfrac{\sqrt{24}}{2\sqrt{6}}$

**15.** _____

**16.** $\dfrac{5\sqrt{48}}{10\sqrt{8}}$

**16.** _____

**17.** $\sqrt{\dfrac{2}{125}} \cdot \sqrt{\dfrac{2}{5}}$

**17.** _____

**18.** $\sqrt{\dfrac{3}{36}} \cdot \sqrt{\dfrac{16}{3}}$

**18.** _____

## Objective 4    Simplify radicals involving variables.

*Simplify each radical. Assume that all variables represent positive real numbers.*

**19.**    $\sqrt{p^2 q^6}$

**19.** _____

**20.**    $\sqrt{48 x^7}$

**20.** _____

**21.**    $\sqrt{32 x^4 y^5}$

**21.** _____

**22.**    $\sqrt{125 x^3 y^4}$

**22.** _____

**23.**    $\sqrt{\dfrac{81}{25 x^6}}$

**23.** _____

**24.** $\sqrt{\dfrac{256y^5}{49x^4}}$

**24.** _____

## Objective 5    Simplify other roots.

*Simplify each expression.*

**25.** $\sqrt[5]{-64}$

**25.** _____

**26.** $\sqrt[3]{\dfrac{64}{27}}$

**26.** _____

**27.** $\sqrt[4]{64}$

**27.** _____

**28.** $\sqrt[3]{\dfrac{1728}{1000}}$

**28.** _____

**29.** $\sqrt[4]{\dfrac{625}{256}}$

**29.** _____

**30.** $\sqrt[4]{3} \cdot \sqrt[4]{27}$

**30.** _____

# Chapter 16 ROOTS AND RADICALS

## 16.3    Adding and Subtracting Radicals

| **Learning Objectives** |
| --- |
| 1    Add and subtract radicals. |
| 2    Simplify radical sums and differences. |
| 3    Simplify more complicated radical expressions. |

## Key Terms

Use the vocabulary terms listed below to complete each statement in exercises 1–3.

**like radicals**          **index**          **unlike radicals**

1.    In the expression, $\sqrt[4]{x^2}$ , the "4" is called the _____.

2.    The expressions $2\sqrt{2}$ and $6\sqrt[3]{2}$ are _____.

3.    The expressions $2\sqrt{2}$ and $7\sqrt{2}$ are _____.

## Objective 1    Add and subtract radicals.

*Add or subtract wherever possible.*

1.    $9\sqrt{11} - 3\sqrt{11}$                              1. _____

2.    $5\sqrt[3]{10} + \sqrt[3]{10}$                         2. _____

3.    $5\sqrt{2} - 4\sqrt{2}$                                3. _____

4.    $5\sqrt{2} + \sqrt{3}$                                 4. _____

5.    $6\sqrt{3} - 2\sqrt{3} + 4\sqrt{3}$                    5. _____

6.    $3\sqrt{5} - 9\sqrt{5} + \sqrt{5}$                     6. _____

**7.** $9\sqrt{2} + 4\sqrt{2} - 7\sqrt{2}$

**7.** _____

**8.** $3\sqrt[4]{12} + 5\sqrt[3]{12} - 8\sqrt{12}$

**8.** _____

**9.** $6\sqrt[5]{4} + 7\sqrt[5]{4} - 9\sqrt[5]{4}$

**9.** _____

**10.** $5\sqrt[3]{3} + 7\sqrt[3]{3} - 9\sqrt{3}$

**10.** _____

**Objective 2    Simplify radical sums and differences.**

*Simplify and add or subtract wherever possible.*

**11.** $4\sqrt{128} + 2\sqrt{32}$

**11.** _____

**12.** $7\sqrt{162} - 9\sqrt{32}$

**12.** _____

**13.** $3\sqrt[3]{16} - 4\sqrt[3]{54}$

**13.** _____

**14.** $6\sqrt{98} + 2\sqrt{72} - 7\sqrt{8}$

**14.** _____

**15.** $5\sqrt{32} - 8\sqrt{18} + 2\sqrt{20}$

**15.** _____

**16.**  $-7\sqrt{63} - 9\sqrt{28} + 6\sqrt{28}$                16. _____

**17.**  $\dfrac{\sqrt{8}}{2} + \dfrac{7}{2}\sqrt{32}$                17. _____

**18.**  $\dfrac{5}{6}\sqrt{72} - \dfrac{3}{2}\sqrt{24}$                18. _____

**19.**  $\sqrt[4]{162} - \sqrt[4]{32}$                19. _____

**20.**  $2\sqrt[4]{243} - 3\sqrt[4]{48}$                20. _____

**Objective 3    Simplify more complicated radical expressions.**

*Perform the indicated operations. Assume that all variables represent nonnegative real numbers.*

**21.**  $\sqrt{5} \cdot \sqrt{7} + 3\sqrt{35}$                21. _____

**22.**  $4\sqrt[3]{r^5} + 3\sqrt[3]{27r^5}$                22. _____

**23.**  $3\sqrt{125x} - \sqrt{80x} + 2\sqrt{45x}$

**23.** _____

**24.**  $4x\sqrt{5} - 3\sqrt{25x} \cdot \sqrt{3x}$

**24.** _____

**25.**  $11\sqrt{5w} \cdot \sqrt{30w} - 8w\sqrt{24}$

**25.** _____

**26.**  $\sqrt{6k^2} \cdot \sqrt{3k} + k\sqrt{2k}$

**26.** _____

**27.**  $3\sqrt{12y} - 2\sqrt{5y} \cdot \sqrt{15}$

**27.** _____

**28.**  $4\sqrt{5r^2} \cdot \sqrt{40} - 7\sqrt{8r^2}$

**28.** _____

**29.**  $3\sqrt{6} \cdot \sqrt{11} - 7\sqrt{66}$

**29.** _____

**30.**  $\frac{1}{3}\sqrt[3]{27x^5 r} + 2x\sqrt[3]{x^2 r} - \frac{1}{2}\sqrt[3]{8x^8 r}$

**30.** _____

# Chapter 16 ROOTS AND RADICALS

## 16.4 Rationalizing the Denominator

**Learning Objectives**

| | |
|---|---|
| 1 | Rationalize denominators with square roots. |
| 2 | Write radicals in simplified form. |
| 3 | Rationalize denominators with cube roots. |

## Key Terms

Use the vocabulary terms listed below to complete each statement in exercises 1–3.

**rationalizing the denominator**     **product rule**     **quotient rule**

1. The _____ states that $\sqrt{a} \cdot \sqrt{b} = \sqrt{ab}$.

2. The process of _____ is changing the denominator of a fraction from a radical to an expression not involving a radical.

3. The _____ states that $\sqrt{\dfrac{a}{b}} = \dfrac{\sqrt{a}}{\sqrt{b}}$.

## Objective 1   Rationalize denominators with square roots.

*Rationalize each denominator.*

1. $\dfrac{-3}{\sqrt{3}}$

1. _____

2. $\dfrac{15}{\sqrt{10}}$

2. _____

3. $\dfrac{-4}{\sqrt{12}}$

3. _____

**4.** $\dfrac{\sqrt{3}}{\sqrt{8}}$

4. _____

**5.** $\dfrac{\sqrt{3}}{\sqrt{24}}$

5. _____

**6.** $\dfrac{6}{\sqrt{28}}$

6. _____

**7.** $\dfrac{\sqrt{4}}{\sqrt{24}}$

7. _____

**8.** $\dfrac{3\sqrt{5}}{\sqrt{125}}$

8. _____

**9.** $\dfrac{25\sqrt{5}}{\sqrt{250}}$

9. _____

**10.**  $\dfrac{2\sqrt{5}}{3\sqrt{75}}$                        10. _____

## Objective 2    Write radicals in simplified form.

*Perform the indicated operations and write all answers in simplest form. Rationalize all denominators. Assume that all variables represent positive real numbers.*

**11.**  $\sqrt{\dfrac{5}{21}} \cdot \sqrt{7}$                    11. _____

**12.**  $\sqrt{\dfrac{3}{2}} \cdot \sqrt{\dfrac{5}{6}}$                   12. _____

**13.**  $\sqrt{\dfrac{63}{50}}$                        13. _____

**14.**  $\sqrt{5} \cdot \sqrt{\dfrac{11}{20}}$                  14. _____

**15.**  $\sqrt{\dfrac{x^6}{y^4}}$                        15. _____

**16.**  $\dfrac{\sqrt{k^2 m^4}}{\sqrt{k^5}}$

**16.** _____

**17.**  $\sqrt{\dfrac{5a^2 b^3}{6}}$

**17.** _____

**18.**  $\sqrt{\dfrac{27}{98}} \cdot \sqrt{\dfrac{1}{3}}$

**18.** _____

**19.**  $\sqrt{\dfrac{72 q t^5}{3 t^6}}$

**19.** _____

**20.**  $\sqrt{\dfrac{20 a^3 b^4}{6 a^2}}$

**20.** _____

## Objective 3    Rationalize denominators with cube roots.

*Rationalize each denominator. Assume that all variables in the denominator represent nonzero real numbers.*

**21.**  $\sqrt[3]{\dfrac{7}{2}}$

**21.** _____

**22.**  $\dfrac{\sqrt[3]{6}}{\sqrt[3]{9}}$

**22.** _____

**23.**  $\dfrac{3}{\sqrt[3]{49}}$

**23.** _____

**24.**  $\sqrt[3]{\dfrac{1}{12}}$

**24.** _____

**25.**  $\sqrt[3]{\dfrac{1}{162}}$

**25.** _____

**26.** $\sqrt[3]{\dfrac{4s}{9r}}$

**26.** _____

**27.** $\sqrt[3]{\dfrac{x^2}{25y}}$

**27.** _____

**28.** $\sqrt[3]{\dfrac{8t}{125u}}$

**28.** _____

**29.** $\sqrt[3]{\dfrac{5}{49x}}$

**29.** _____

**30.** $\sqrt[3]{\dfrac{7x^2}{81y^2}}$

**30.** _____

# Chapter 16 ROOTS AND RADICALS

## 16.5   More Simplifying and Operations with Radicals

| **Learning Objectives** |
| --- |
| 1      Simplify products of radical expressions. |
| 2      Use conjugates to rationalize denominators of radical expressions. |
| 3      Write radical expressions with quotients in lowest terms. |

### Key Terms

Use the vocabulary terms listed below to complete each statement in exercises 1–2.

**conjugate      rationalize the denominator**

1.   To _____ of $\dfrac{3}{\sqrt{5}}$, multiply both the numerator

     and the denominator by $\sqrt{5}$.

2.   The _____ of $a + b$ is $a - b$.

### Objective 1   Simplify products of radical expressions.

*Find each product, and simplify.*

1.   $\sqrt{5}\left(\sqrt{12} + 4\sqrt{7}\right)$                              1. _____

2.   $\sqrt{7}\left(2\sqrt{8} - 9\sqrt{7}\right)$                              2. _____

3.   $\left(4\sqrt{5} + \sqrt{3}\right)\left(\sqrt{2} - \sqrt{7}\right)$                  3. _____

**4.** $\left(\sqrt{5} - \sqrt{8}\right)\left(\sqrt{3} + \sqrt{2}\right)$

4. _____

**5.** $\left(\sqrt{6} - 2\sqrt{5}\right)\left(4\sqrt{6} + \sqrt{10}\right)$

5. _____

**6.** $\left(4\sqrt{2} + 3\sqrt{3}\right)\left(\sqrt{2} - 7\sqrt{3}\right)$

6. _____

**7.** $\left(2\sqrt{3} - 5\sqrt{2}\right)\left(2\sqrt{3} + 5\sqrt{2}\right)$

7. _____

**8.** $\left(\sqrt{5} - \sqrt{8}\right)^2$

8. _____

**9.** $\left(7 + 4\sqrt{3}\right)^2$

9. _____

**10.** $\left(3\sqrt{2} - 2\sqrt{3}\right)^2$

10. _____

**Objective 2    Use conjugates to rationalize denominators of radical expressions.**

*Rationalize each denominator. Write quotients in lowest terms.*

**11.**    $\dfrac{4}{\sqrt{3}+2}$                       **11.** _____

**12.**    $\dfrac{\sqrt{2}}{\sqrt{5}-2}$                     **12.** _____

**13.**    $\dfrac{3}{\sqrt{2}-\sqrt{5}}$                    **13.** _____

**14.**    $\dfrac{4}{\sqrt{5}+\sqrt{2}}$                    **14.** _____

**15.**    $\dfrac{5}{\sqrt{3}-\sqrt{10}}$                   **15.** _____

**16.**   $\dfrac{\sqrt{3}+\sqrt{2}}{\sqrt{3}-\sqrt{2}}$             **16.** _____

**17.**   $\dfrac{\sqrt{6}+2}{\sqrt{2}-4}$              **17.** _____

**18.**   $\dfrac{5\sqrt{5}}{4-\sqrt{15}}$             **18.** _____

**19.**   $\dfrac{\sqrt{5}-2}{\sqrt{3}+2}$             **19.** _____

**20.**   $\dfrac{\sqrt{2}+\sqrt{3}}{\sqrt{2}-\sqrt{3}}$             **20.** _____

## Objective 3 Write radical expressions with quotients in lowest terms.

*Write each quotient in lowest terms.*

21. $\dfrac{2\sqrt{7} - 4\sqrt{2}}{6}$

21. _____

22. $\dfrac{8 + 6\sqrt{7}}{4}$

22. _____

23. $\dfrac{4\sqrt{2} - 6}{6}$

23. _____

24. $\dfrac{6 + \sqrt{8}}{2}$

24. _____

25. $\dfrac{3 + \sqrt{27}}{9}$

25. _____

**26.**  $\dfrac{8\sqrt{5}-12}{20}$

**26.** _____

**27.**  $\dfrac{\sqrt{5}+4}{\sqrt{3}-\sqrt{2}}$

**27.** _____

**28.**  $\dfrac{12+6\sqrt{6}}{8}$

**28.** _____

**29.**  $\dfrac{135\sqrt{3}+25}{5}$

**29.** _____

**30.**  $\dfrac{72\sqrt{2}-16\sqrt{7}}{24}$

**30.** _____

# Chapter 16 ROOTS AND RADICALS

## 16.6  Solving Equations with Radicals

| **Learning Objectives** |
| --- |
| 1      Solve radical equations having square root radicals. |
| 2      Identify equations with no solutions. |
| 3      Solve equations by squaring a binomial. |
| 4      Solve problems using formulas that involve radicals. |

## Key Terms

Use the vocabulary terms listed below to complete each statement in exercises 1–2.

         **radical equation**             **extraneous solution**

1.    An _____ is a potential solution to an equation that does not satisfy the equation..

2.    An equation with a variable in the radicand is a _____.

## Objective 1  Solve radical equations having square root radicals.

*Solve each equation.*

1.    $\sqrt{3x+1} = 3$                  1. _____

2.    $2\sqrt{r} = \sqrt{3r+16}$            2. _____

3.    $\sqrt{2+4k} = 3\sqrt{k}$            3. _____

**4.**     $\sqrt{6r+4} = 3\sqrt{r}$

**5.**     $\sqrt{3x+3} = 2\sqrt{3x}$

**6.**     $\sqrt{x-3} - 8 = 0$

**7.**     $\sqrt{4x+3} = \sqrt{3x+5}$

**8.**     $\sqrt{5y-2} = \sqrt{2y+1}$

**9.**     $\sqrt{4x+18} = \sqrt{3x+15}$

## Objective 2    Identify equations with no solutions.

*Solve each equation.*

**10.**    $\sqrt{y+3} = -4$                   **10.** _____

**11.**    $\sqrt{x+2} + 7 = 0$             **11.** _____

**12.**    $\sqrt{x+1} + 9 = 0$             **12.** _____

**13.**    $\sqrt{2m+3} = 3\sqrt{m+5}$        **13.** _____

**14.**    $\sqrt{x-7} = \sqrt{2x+5}$          **14.** _____

**15.**    $\sqrt{4-3n} = 2\sqrt{-5n}$

**16.**    $r = \sqrt{r^2 - 6r + 12}$

**17.**    $s = \sqrt{s^2 + 4s + 4}$

**18.**    $\sqrt{d^2 + 4d + 12} + d = 0$

**Objective 3    Solve equations by squaring a binomial.**

*Solve each equation.*

**19.**    $\sqrt{b-4} = b - 6$

**20.**   $\sqrt{2x-1} = x - 2$

**20.** _____

**21.**   $q - 1 = \sqrt{q^2 - 4q + 7}$

**21.** _____

**22.**   $3\sqrt{p+6} = p + 6$

**22.** _____

**23.**   $2\sqrt{c+2} = c + 3$

**23.** _____

**24.**   $\sqrt{5x+9} = 2x - 3$

**24.** _____

**25.**   $\sqrt{2y+6} = y - 1$

**25.** _____

**26.** $p = \sqrt{p^2 - 3p + 12}$

**27.** $3t = \sqrt{9t^2 + t - 4}$

**Objective 4    Solve problems using formulas that involve radicals.**

*Use Heron's formula* $= \sqrt{s(s-a)(s-b)(s-c)}$ *to find the area of the triangle with the given sides. Round answers to the nearest whole number.*

**28.** $a = 10$ in., $b = 10$ in., $c = 12$ in.

**29.** $a = 42.3$ cm, $b = 29.8$ cm, $c = 33.7$ cm

**30.** $a = 23$ m, $b = 19$ m, $c = 12$ m

# Chapter 17 QUADRATIC EQUATIONS

### 17.1    Solving Quadratic Equations by the Square Root Property

| **Learning Objectives** |
| --- |
| 1      Solve equations of the form $x^2 = k$, where $k > 0$. |
| 2      Solve equations of the form $(ax + b)^2 = k$, where $k > 0$. |
| 3      Use formulas involving squared variables. |

### Key Terms

Use the vocabulary terms listed below to complete each statement in exercises 1–2.

**quadratic equation**          **zero-factor property**

1.    An equation that can be written in the form $ax^2 + bx + c = 0$ is a

   _____.

2.    The _____ states that if a product equals 0, then at least one of the factors of the product also equals zero.

### Objective 1    Solve equations of the form $x^2 = k$, where $k > 0$.

*Solve each equation by using the square root property. Express all radicals in simplest form.*

1.    $r^2 = 900$                                    1. _____

2.    $d^2 - 250 = 0$                              2. _____

3.    $c^2 + 36 = 0$                                3. _____

4.    $t^2 - 12.25 = 0$                            4. _____

**5.** $m^2 = 1.96$

**5.** _____

**6.** $121x^2 - 24 = 0$

**6.** _____

**7.** $t^2 - 30.25 = 0$

**7.** _____

**8.** $s^2 - 98 = 0$

**8.** _____

**9.** $289h^2 - 90 = 0$

**9.** _____

**10.** $p^2 = -144$

**10.** _____

**Objective 2** **Solve equations of the form** $(ax + b)^2 = k$, **where** $k > 0$.

*Solve each equation by using the square root property. Express all radicals in simplest form.*

**11.** $(y + 2)^2 = 16$

**11.** _____

**12.**   $(p - 9)^2 = 28$

**12.** _____

**13.**   $(7p - 4)^2 = 289$

**13.** _____

**14.**   $(2x + 5)^2 = 32$

**14.** _____

**15.**   $(q - 4)^2 - 7 = 0$

**15.** _____

**16.**   $(6p + 9)^2 = 54$

**16.** _____

**17.**   $(10m - 5)^2 - 9 = 0$

**17.** _____

**18.**    $(3f + 4)^2 = 32$

**18.** _____

**19.**    $\left(\frac{1}{2}z + 4\right)^2 = 81$

**19.** _____

**20.**    $\left(\frac{1}{4}x - 2\right)^2 = 49$

**20.** _____

**21.**    $(10m - 5)^2 + 9 = 0$

**21.** _____

**22.**    $\left(\frac{1}{3}r - 3\right)^2 = 50$

**22.** _____

## Objective 3    Use formulas involving squared variables.

*Write each rational expression in lowest terms. Assume that no values of any variable make any denominator zero.*

23.    The distance *d* traveled by a freely falling object in *t* seconds is given by the formula $d = 16t^2$. How long will it take for an object to fall 1500 feet?

23. _____

24.    The formula $A = P(1 + r)^2$ gives the amount *A* that *P* dollars invested at an annual rate of interest *r* will grow to in two years. Mary invests $1500, and after two years, she has $1653.75. What is the interest rate?

24. _____

25.    The volume of a cylinder is given by the formula $V = \pi r^2 h$, where *V* = the volume, *r* = the radius of the base of the cylinder, and *h* = the height of the cylinder. If the volume of a can is $20\pi$ in.$^3$, and its height is 5 inches, find the radius of the can.

25. _____

26.    The volume of a cone is given by the formula $V = \dfrac{1}{3}\pi r^2 h$, where *V* = the volume, *r* = the radius of the base of the cone, and *h* = the height of the cone. If the volume of a cone is $15\pi$ cm.$^3$, and its height is 9 cm, find the radius of the cone.

26. _____

**27.** The length of a room is twice the width. If the area of the room is 200 ft$^2$, find the dimensions of the room.

**27.** width _____

length _____

**28.** If the area of a square is doubled, the result is 72 m$^2$. Find the length of each side of the square.

**28.** _____

**29.** One leg of a right triangle has length 5 cm, and the hypotenuse has length 10 cm. Find the length of the other leg. (Hint: Use the Pythagorean theorem.)

**29.** _____

**30.** If the length of each side of a square is doubled, the area of the resulting square is 72 m$^2$. Find the length of each side of the original square.

**30.** _____

# Chapter 17 QUADRATIC EQUATIONS

## 17.2 Solving Quadratic Equations by Completing the Square

| Learning Objectives |
| --- |
| 1     Solve quadratic equations by completing the square when the coefficient of the second-degree term is 1. |
| 2     Solve quadratic equations by completing the square when the coefficient of the second-degree term is not 1. |
| 3     Simplify the terms of an equation before solving. |
| 4     Solve applied problems that require quadratic equations. |

## Key Terms

Use the vocabulary terms listed below to complete each statement in exercises 1–3.

**completing the square**     **perfect square trinomial**     **square root property**

1.  A _____ can be written in the form $x^2 + 2kx + k^2$ or $x^2 - 2kx + k^2$

2.  The _____ says that, if $k$ is positive and $a^2 = k$, then $a = \pm\sqrt{k}$.

3.  Use the process called _____ in order to rewrite an equation so it can be solved using the square root property.

## Objective 1    Solve quadratic equations by completing the square when the coefficient of the second-degree term is 1.

*Solve each equation by completing the square.*

1.  $x^2 + 3x = 4$

    1. _____

2.  $r^2 + 8r = -4$

    2. _____

3.  $x^2 - 4x = 2$

    3. _____

**4.**    $w^2 - 3w - 4 = 0$                          **4.** _____

**5.**    $x^2 + 2x = 63$                             **5.** _____

**6.**    $d^2 + 10d - 11 = 0$                        **6.** _____

**7.**    $c^2 - c - \frac{5}{2} = 0$                 **7.** _____

**Objective 2    Solve quadratic equations by completing the square when the coefficient of the second-degree term is not 1.**

*Solve each equation by completing the square.*

**8.**    $3m^2 - 15m = 42$                           **8.** _____

**9.**    $3r^2 = 6r + 2$                             **9.** _____

**10.**   $2p^2 + 6p - 1 = 0$                         **10.** _____

**11.**   $6x^2 - x = 15$                             **11.** _____

**12.**  $3x^2 - 2x + 4 = 0$

**12.** _____

**13.**  $-y^2 - 2y + 8 = 0$

**13.** _____

**14.**  $3t^2 + t - 2 = 0$

**14.** _____

**15.**  $6q^2 + 4q = 1$

**15.** _____

**Objective 3    Simplify the terms of an equation before solving.**

_Simplify each of the following equations and then solve by completing the square._

**16.**  $2x - 4 = x^2 - 2x$

**16.** _____

**17.**  $4y^2 + 6y = 2y + 3$

**17.** _____

**18.**  $6y^2 + 3y = 4y^2 + y - 5$

**18.** _____

**19.**    $2z^2 = 6z + 3 - 4z^2$          **19.** _____

**20.**    $(b-1)(b+7) = 9$          **20.** _____

**21.**    $(s+3)(s+1) = 1$          **21.** _____

**22.**    $(j+3)(j-2) = 5$          **22.** _____

**Objective 4    Solve applied problems that require quadratic equations.**

*Divide. Write each answer in lowest terms.*

**23.**    A certain projectile is located at a distance of $d = 3t^2 - 6t + 1$ feet from its starting point after $t$ seconds. How many seconds will it take the projectile to travel 10 feet?          **23.** _____

**24.**  The time $t$ in seconds for a car to skid 48 feet is given (approximately) by $48 = 64t - 16t^2$. Solve this equation for $t$. Are both answers reasonable?

**24.** _____

**25.**  The amount $A$ that $P$ dollars invested at a rate of interest $r$ will amount to in two years is $A = P(1+r)^2$. At what interest rate will \$100 grow to \$110.25 in two years?

**25.** _____

**26.**  If Dipti throws a ball into the air from ground level with an initial velocity of 32 feet per second, its height $s$ (in feet) after $t$ seconds is given by the formula $s = -16t^2 + 32t$. After how many seconds will the ball return to the ground?

**26.** _____

**27.**  If James throws an object upward from ground level with an initial velocity of 80 feet per second, its height $s$ (in feet) after $t$ seconds is given by the formula $s = -16t^2 + 80t$. After how many seconds will the object reach a height of 64 feet?

**27.** _____

**28.** A rule for estimating the number of board feet of lumber that can be cut from a log depends on the diameter and length of the log. To find the diameter $d$ (in inches) needed to get $x$ board feet of lumber from an 8-foot log, use the formula, $\left(\dfrac{d-4}{4}\right)^2 = x$.

Find the diameter needed to get 20 board feet of lumber.

**28.**

**29.** The commodities market is very unstable; money can be made or lost quickly on investments in soybeans, wheat, pork bellies, and so on. Suppose that an investor kept track of his total profit, $P$ (in thousands of dollars), at time $t$ (in months), after he began investing, and found that his profit was given by the formula $P = 4t^2 - 24t + 32$. Find the times at which he broke even on his investment.

**29.**

**30.** George and Albert have found that the profit (in dollars) from their cigar shop is given by the formula $P = -10x^2 + 100x + 300$, where $x$ is the number of units of cigars sold daily. How many units should be sold for a profit of $460?

**30.** _____

# Chapter 17 QUADRATIC EQUATIONS

## 17.3   Solving Quadratic Equations by the Quadratic Formula

| **Learning Objectives** |
| --- |
| 1    Identify the values of $a$, $b$, and $c$ in a quadratic equation. |
| 2    Use the quadratic formula to solve quadratic equations. |
| 3    Solve quadratic equations with only one solution. |
| 4    Solve quadratic equations with fractions. |

### Key Terms

Use the vocabulary terms listed below to complete each statement in exercises 1–3.

**quadratic formula**          **standard form**          **constant**

1.    A quadratic equation written in the form $ax^2 + bx + c = 0$, $a \neq 0$ is written in

_____.

2.    A symbol that represents a value that doesn't change is a _____.

3.    The formula $x = \dfrac{-b \pm \sqrt{b^2 - 4ac}}{2a}$ is called the _____.

### Objective 1    Identify the values of $a$, $b$, and $c$ in a quadratic equation.

*Write each equation in standard form, if necessary, and then identify the values of a, b, and c. Do not actually solve the equation.*

1.    $10x^2 = -4x$

1. _____

2.    $4p = -4p^2 + 7$

2. _____

3.    $3p^2 = 12$

3. _____

4.    $(z+1)(z+2) = -7$

4. _____

**5.**   $3 - \left[ (r+2)(r-3) \right] = 2r$

**5.** _____

**Objective 2**   **Use the quadratic formula to solve quadratic equations.**

*Use the quadratic formula to solve each equation. Write all radicals in simplified form, and write all answers in lowest terms.*

**6.**   $2k^2 - 2k - 5 = 0$

**6.** _____

**7.**   $y^2 = 13 - 12y$

**7.** _____

**8.**   $x^2 - 3x + 1 = 8$

**8.** _____

**9.**   $-7r^2 = 5r + 3$

**9.** _____

**10.**   $(x-4)(x+3) = 8$

**10.** _____

**11.**   $5k^2 + 4k - 2 = 0$                    **11.** _____

**12.**   $3y(4y - 3) = 4y - 3$                  **12.** _____

**13.**   $3p(p + 2) = 3p^2 + 7$                 **13.** _____

## Objective 3   Solve quadratic equations with only one solution.

*Use the quadratic formula to solve each equation. Write all radicals in simplified form, and write all answers in lowest terms.*

**14.**   $16a^2 - 8a + 1 = 0$                   **14.** _____

**15.**   $m^2 + 4 = 4m$                         **15.** _____

**16.**   $4q^2 + 12q + 9 = 0$                   **16.** _____

**17.** $9r^2 = 6r - 1$ 

_____

**18.** $49a^2 - 126a = -81$ 

_____

**19.** $n^2 + 81 = 18n$ 

_____

**20.** $d^2 = 20d - 100$ 

_____

**21.** $100p^2 + 60p + 9 = 0$ 

_____

**22.** $144t^2 + 168t + 49 = 0$ 

_____

## Objective 4   Solve quadratic equations with fractions.

*Use the quadratic formula to solve each equation. Write all radicals in simplified form, and write all answers in lowest terms.*

**23.**   $-\frac{1}{4}x^2 + 4 = \frac{1}{2}x$

**23.** _____

**24.**   $\frac{1}{6}y^2 + \frac{1}{2}y = \frac{2}{3}$

**24.** _____

**25.**   $\frac{1}{4}t^2 - \frac{1}{3}t + \frac{5}{12} = 0$

**25.** _____

**26.**   $k^2 + \frac{8}{3}k - \frac{1}{3} = 0$

**26.** _____

**27.** $\frac{1}{3}x^2 + \frac{2}{3}x = 1$

27. _____

**28.** $\frac{1}{16}z^2 + \frac{3}{8}z = -\frac{1}{2}$

28. _____

**29.** $\frac{1}{20}z^2 = \frac{1}{5}z + \frac{1}{8} - \frac{1}{10}z$

29. _____

**30.** $\frac{1}{2}x^2 + \frac{1}{4}x - \frac{15}{4} = 0$

30. _____

# Chapter 17 QUADRATIC EQUATIONS

## 17.4    Graphing Quadratic Equations

| Learning Objectives |
| --- |
| 1      Graph quadratic equations. |
| 2      Find the vertex of a parabola. |

### Key Terms

Use the vocabulary terms listed below to complete each statement in exercises 1–4.

**parabola**      **vertex**      **axis**      **line of symmetry**

1. If a graph is folded on its_____, the two sides coincide.

2. The _____ of a parabola that opens upward or downward is the lowest or highest point on the graph.

3. The _____ of a parabola that opens upward or downward is a vertical line through the vertex.

4. The graph of the quadratic equation $y = ax^2 + bx + c$ is called a

     _____.

**Objective 1**    **Graph quadratic equations.**

**Objective 2**    **Find the vertex of a parabola.**

*Graph each equation. Give the coordinates of the vertex in each case.*

1.    $y = -x^2$

**1.**

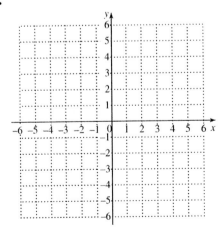

vertex: _____

**2.** $y = -x^2 - 1$

**2.**

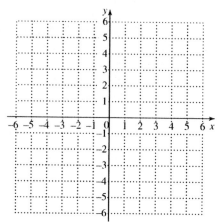

vertex: _____

**3.** $y = x^2 - 3$

**3.**

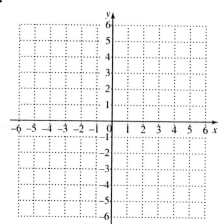

vertex: _____

**4.** $y = \frac{1}{3}x^2$

**4.**

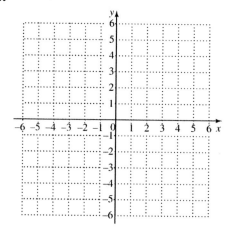

vertex: _____

**5.**   $y = x^2 + 2$

**5.**

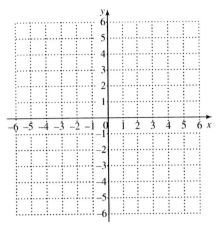

vertex: _____

**6.**   $y = 2x^2 - 4$

**6.**

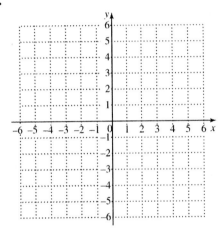

vertex: _____

**7.**   $y = 1 - 2x^2$

**7.**

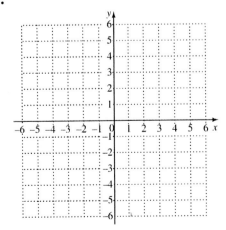

vertex: _____

**8.** $y = 9 - x^2$

**8.**

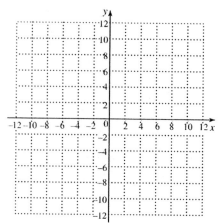

vertex: _____

**9.** $y = (x - 2)^2$

**9.**

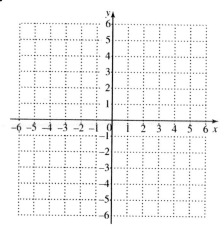

vertex: _____

**10.** $y = (x + 4)^2$

**10.**

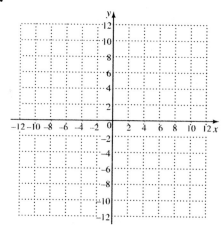

vertex: _____

**11.**   $y = (x - 3)^2$

**11.**

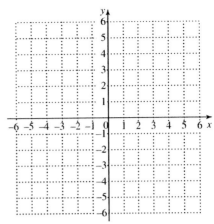

vertex: _____

**12.**   $y = -(x + 1)^2$

**12.**

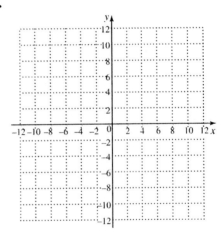

vertex: _____

**13.**   $y = 2x^2 + 4x$

**13.**

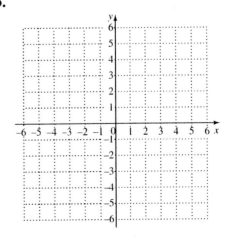

vertex: _____

**14.** $y = x^2 + x - 2$

**14.**

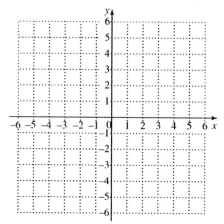

vertex: _____

**15.** $y = x^2 - 6x + 11$

**15.**

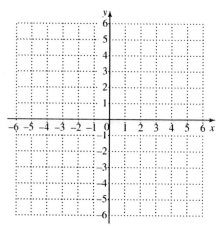

vertex: _____

**16.** $y = -x^2 + 6x - 9$

**16.**

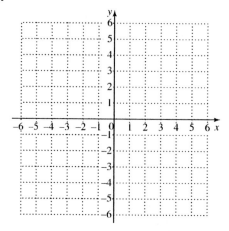

vertex: _____

**17.**    $y = -x^2 + 4x - 1$

**17.**

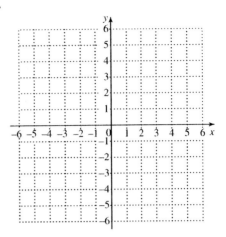

vertex: _____

**18.**    $y = x^2 + 8x + 14$

**18.**

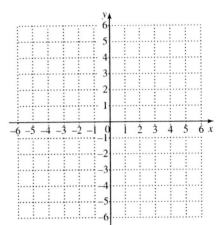

vertex: _____

**19.**    $y = -x^2 + 6x - 13$

**19.**

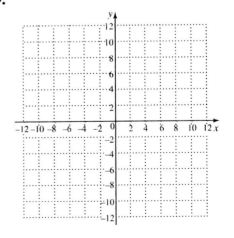

vertex: _____

**20.**   $y = x^2 + 2x - 2$

**20.**

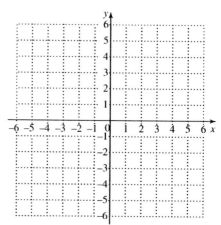

vertex: _____

**21.**   $y = -x^2 + 5x$

**21.**

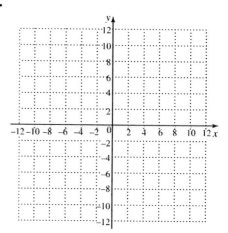

vertex: _____

**22.**   $y = -x^2 - 3x + 1$

**22.**

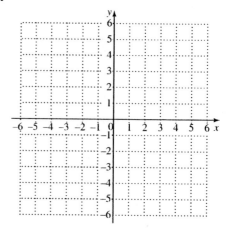

vertex: _____

**23.**    $y = x^2 - 6x + 9$

**23.**

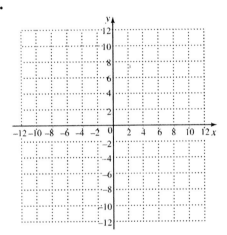

vertex: _____

**24.**    $y = x^2 + 4x - 5$

**24.**

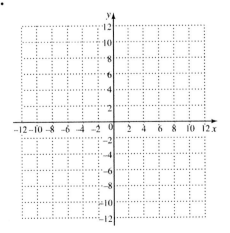

vertex: _____

**25.**    $y = -x^2 - 2x + 8$

**25.**

vertex: _____

**26.** $y = \frac{1}{2}(x+2)^2$

**26.**

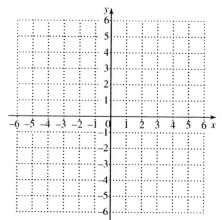

vertex: _____

**27.** $y = x^2 - 2x - 3$

**27.**

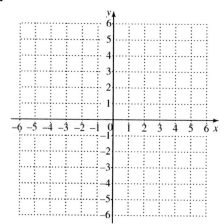

vertex: _____

**28.** $y = (x-2)(x+4)$

**28.**

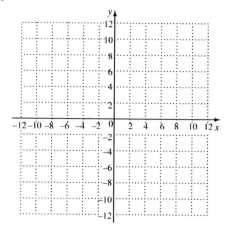

vertex: _____

**29.**    $y = -\frac{1}{2}x^2 + 2x$

**29.**

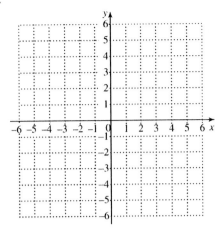

vertex: _____

**30.**    $y = \frac{1}{3}x^2 + x - 1$

**30.**

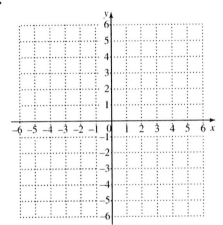

vertex: _____

# Chapter 17 QUADRATIC EQUATIONS

## 17.5    Introduction to Functions

| **Learning Objectives** |
| --- |
| 1    Understand the definition of a relation. |
| 2    Understand the definition of a function. |
| 3    Decide whether an equation defines a function. |
| 4    Use function notation. |
| 5    Apply the function concept in an application. |

### Key Terms

Use the vocabulary terms listed below to complete each statement in exercises 1–5.

**components**        **relation**        **domain**        **range**        **function**

1.    Any set of ordered pairs is called a _____.

2.    The set of all second components in the ordered pairs of a relation is the _____ of the relation.

3.    A _____ is a set of ordered pairs in which each first component corresponds to exactly one second component.

4.    In an ordered pair $(x, y)$, $x$ and $y$ are the _____.

5.    The set of all first components in the ordered pairs of a relation is the _____ of the relation.

**Objective 1    Understand the definition of a relation.**

**Objective 2    Understand the definition of a function.**

*Decide whether or not the relations graphed or defined are functions and give the domain and range of each.*

1.    $\{(2,7),(5,-4),(-3,-1),(0,-8),(5,2)\}$

1. _____

domain: _____

range: _____

2.    $\{(1,3),(5,7),(11,9),(8,-2),(6,-7),(-4,-3)\}$

2. _____

domain: _____

range: _____

**3.**  $\{(0,1),(2,6),(-3,7),(2,9),(-7,1),(-3,4)\}$

**3.** _____

domain: _____

range: _____

**4.**  $\{(3,5),(3,8),(3,-4),(3,1),(3,0)\}$

**4.** _____

domain: _____

range: _____

**5.**  $\{(1,4),(3,4),(7,4),(-2,4),(-5,4)\}$

**5.** _____

domain: _____

range: _____

**6.**  $\{(-1.2,4),(1.8,-2.5),(3.7,-3.8),(3.7,3.8)\}$

**6.** _____

domain: _____

range: _____

**7.**  $\{(-3,5),(-2,5),(-1,0),(0,-5),(1,5)\}$

**7.** _____

domain: _____

range: _____

**8.**

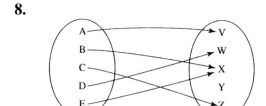

**8.** _____

domain: _____

range: _____

**9.**

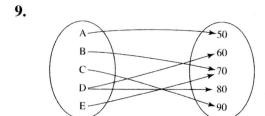

**9.** _____

domain: _____

range: _____

*Use the vertical line test to determine whether each relation graphed is a function.*

**10.**

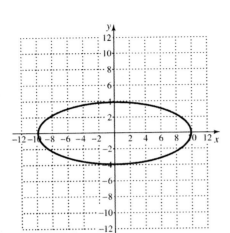

**10.** _____

**11.**

**11.** _____

**12.**

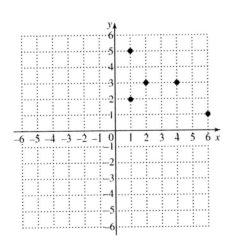

**12.** _____

## Objective 3    Decide whether an equation defines a function.

*Decide whether each equation defines y as a function of x.*

**13.**    $y = 3x + 2$

**13.** _____

**14.**    $y = 7$

**14.** _____

**15.**    $x = -4$

**15.** _____

**16.**    $x^2 + y^2 = 4$

**16.** _____

**17.**    $3x + 2y = 8$

**17.** _____

**18.**    $x = y^2 - 4$

**18.** _____

## Objective 4    Use function notation.

*For each function f, find (a) $f(-2)$, (b) $f(0)$, and (c) $f(4)$.*

**19.**    $f(x) = 3x - 7$

**19. a.** _____

**b.** _____

**c.** _____

**20.**    $f(x) = x^2 - 3x + 2$

**20. a.** _____

**b.** _____

**c.** _____

**21.**   $f(x) = 2x^2 + x - 5$

**21. a.** _____

**b.** _____

**c.** _____

**22.**   $f(x) = |2x + 3|$

**22. a.** _____

**b.** _____

**c.** _____

**23.**   $f(x) = x^3 - 2x^2 + 4$

**23. a.** _____

**b.** _____

**c.** _____

**24.**   $f(x) = 9$

**24. a.** _____

**b.** _____

**c.** _____

## Objective 5   Apply the function concept in an application.

*The number of internet users in the world is shown in the graph below. Use the graph to answer problems 25–28.*

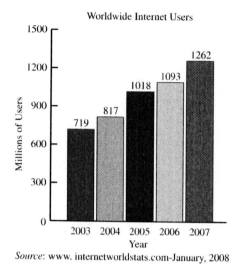

**Worldwide Internet Users**

*Source*: www. internetworldstats.com-January, 2008

**25.**   Write the information in the graph as a set of ordered     **25.** _____
pairs. Does this set define a function?

**26.**   Suppose that *w* is the name given to this relation.     **26.** domain _____
Give the domain and range of *w*.

range _____

**27.**   Find *w*(2003) and *w*(2006).     **27.** _____

**28.**   For what value of *x* does *w*(*x*) = 1018 (million)?     **28.** _____

*The yearly revenue for a small business is shown in the table below. Use the table to answer problems 29–30.*

| Year | Revenue (thousands of dollars) |
|------|-------------------------------|
| 2001 | 596 |
| 2002 | 625 |
| 2003 | 872 |
| 2004 | 795 |
| 2005 | 625 |

**29.** Write the information in the graph as a set of ordered pairs. Does this set define a function?

**29.** _____

**30.** Suppose that *r* is the name given to this relation. Give the domain and range of *r*.

**30.** domain _____

range _____

# Chapter 1 WHOLE NUMBERS

## 1.1 Reading and Writing Whole Numbers

### Key Terms

1. table  2. whole numbers  3. place value

### Objective 1

1. whole number  3. not a whole number

### Objective 2

5. 9; 4  7. 1; 1  9. 2; 0

11. 75; 229; 301  13. 300; 459; 200; 5

### Objective 3

15. thirty-nine thousand, fifteen

17. two million, fifteen thousand, one hundred two

19. 4127  21. 685,000,259  23. 7210

25. 15,313

### Objective 4

27. 177  29. 315

## 1.2 Adding Whole Numbers

### Key Terms

1. commutative property of addition
2. addends
3. addition
4. associative property of addition
5. regrouping (carrying)
6. perimeter
7. sum

### Objective 1

1. 12

### Objective 2

3. 25      5. 31

### Objective 3

7. 99      9. 98,977

### Objective 4

11. 112      13. 15,815      15. 4322

### Objective 5

17. 38 miles      19. 44 miles      21. 625 tickets

23. 310 feet      25. 1044 yards

### Objective 5

27. incorrect; 17,280   29. correct

## 1.3    Subtracting Whole Numbers

### Key Terms

1. minuend
2. regrouping (borrowing)
3. subtrahend
4. difference

### Objective 1

1. $187 - 38 = 149$; $187 - 149 = 38$

3. $785 + 426 = 1211$

### Objective 2

5. minuend: 98; subtrahend: 38; difference: 62

### Objective 3

7. 5151

### Objective 4

9. not correct; 153    11. not correct; 2980    13. not correct; 78,087

### Objective 5

15. 192    17. 25,899    19. 245

### Objective 6

21. 25 boxes    23. $419    25. $5184

27. 2758 people    29. 76 athletes

## 1.4    Multiplying Whole Numbers

### Key Terms

1.  commutative property of multiplication
2.  factors
3.  associative property of multiplication
4.  multiple
5.  product
6.  chain multiplication problem

### Objective 1

1.  factors: 5, 2; product: 10

### Objective 2

3.  32
5.  192

### Objective 3

7.  815
9.  285,867

### Objective 4

11.  439,000
13.  6,010,000
15.  16,468,000

### Objective 5

17.  193,488
19.  330,687

### Objective 6

21.  560 yards
23.  $912
25.  576 miles
27.  $9728
29.  $1602

## 1.5 Dividing Whole Numbers

**Key Terms**

1. remainder
2. dividend
3. quotient
4. divisor
5. short division

**Objective 1**

1. $3\overline{)15}^{\,5}$; $\dfrac{15}{3} = 5$

**Objective 2**

3. dividend: 63; divisor: 7; quotient: 9

5. dividend: 44; divisor: 11; quotient: 4

**Objective 3, Objective 4**

7. undefined
9. undefined
11. undefined

**Objective 5, Objective 6**

13. 38

**Objective 7**

15. 144 R 4
17. 141 R 7
19. 170 R 2

**Objective 8**

21. correct
23. incorrect; 5814
25. correct

**Objective 9**

27. 2: no; 3: yes; 5: no; 10: no
29. 2: no; 3: no; 5: yes; 10: no

## 1.6　Long Division

### Key Terms

1. dividend
2. remainder
3. divisor
4. quotient
5. long division

### Objective 1

1. 82
3. 77
5. 102
7. 85 R84
9. 309 R1
11. 6354 R7
13. 32,587
15. 654 R22

### Objective 2

17. 7
19. 6
21. 42
23. 800

### Objective 3

25. correct
27. 45 R23
29. incorrect; 296 R79

### 1.7    Rounding Whole Numbers

**Key Terms**

1. front end rounding   2. rounding   3. estimate

**Objective 1**

1. 257,3_0_1          3. 6_4_5,371

**Objective 2**

5. 7900          7. 18,200          9. 8400

11. 52,000          13. 53,600          15. 600,000

**Objective 3**

17. $40 + 20 + 60 + 90 = 210$; 210

19. $70 - 40 = 30$; 27

21. $300 + 300 + 200 + 900 = 1700$; 1698

23. $1000 - 400 = 600$; 589

25. $900 \times 800 = 720,000$; 715,008

27. $600 + 40 + 200 + 2000 = 2840$; 3280

29. $1000 \times 40 = 40,000$; 36,260

## 1.8 Exponents, Roots, and Order of Operations

### Key Terms

1. order of operations

2. square root

3. perfect square

## Objective 1

1. exponent: 2; base: 7; 49

3. exponent: 3; base: 8; 512

## Objective 2

5. 4

7. 11

9. 15

11. 2500; 2500

13. 400; 20

## Objective 3

15. 39

17. undefined

19. 27

21. 54

23. 45

25. 75

27. 28

29. 45

**1.9    Reading Pictographs, Bar Graphs, and Line Graphs**

**Key Terms**

1.    line graph        2.    pictograph        3.    bar graph

**Objective 1**

1.    Georgia        3.    Minnesota        5.    1%

7.    3              9.    Spanish

**Objective 2**

11.    10        13.    4        15.    18

17.    2900        19.    freshman

**Objective 3**

21.    The net sales are increasing every year.        23.    2007

25.    2003–2004        27.    2005        29.    $1.5 million

Answers to Worksheets for Classroom or Lab Practice

## 1.10    Solving Application Problems

### Key Terms

1. indicator words    2. sum; increased by    3. product; times

4. quotient; per    5. difference; fewer

### Objective 1

1. subtraction    3. multiplication    5. subtraction

### Objective 2

7. 4 strips    9. 1011 vehicles    11. 6850 people

13. 589 deer    15. 80,128,450 gallons    17. $3826

19. $1854

### Objective 3

21. $600 \div 60 = 10$ hours; 11 hours

23. $40 \times 30 = 1200$ miles; 936 miles

25. $40,000 \times 10 = 400,000$ square feet; 217,800 square feet

27. $(20 \times 10) \div 10 \times \$20 = \$400$; $552

29. $300 - (40 + 20 + 80) + (20 + 40 + 100) = 320$ machines; 348 machines

## Chapter 2 MULTIPLYING AND DIVIDING FRACTIONS

### 2.1    Basics of Fractions

### Key Terms

1.  improper fraction    2.  numerator        3.  proper fraction

4.  denominator

**Objective 1**  In each answer, the first fraction is the shaded portion, and the second fraction is the unshaded portion.

1.  $\dfrac{3}{8}; \dfrac{5}{8}$        3.  $\dfrac{5}{6}; \dfrac{1}{6}$        5.  $\dfrac{5}{8}; \dfrac{3}{8}$

7.  $\dfrac{8}{5}; \dfrac{2}{5}$        9.  $\dfrac{7}{10}; \dfrac{3}{10}$

### Objective 2

11.  N: 4; D: 3        13.  N: 8; D:11        15.  N: 19; D:50

17.  N: 19; D: 8        19.  N: 157; D: 12

### Objective 3

21.  improper        23.  proper        25.  proper

27.  improper        29.  proper

## 2.2 Mixed Numbers

### Key Terms

1.  proper fraction
2.  mixed number
3.  improper fraction
4.  whole numbers

### Objective 1

1.  $2\frac{1}{2}, 1\frac{1}{6}$
3.  none
5.  $4\frac{3}{4}$

### Objective 2

7.  $\frac{11}{6}$
9.  $\frac{39}{7}$
11.  $\frac{25}{4}$
13.  $\frac{15}{2}$
15.  $\frac{38}{7}$
17.  $\frac{79}{9}$

### Objective 3

19.  $1\frac{5}{9}$
21.  $3\frac{2}{9}$
23.  $4\frac{1}{5}$
25.  $2\frac{7}{9}$
27.  $30\frac{2}{3}$
29.  $44\frac{1}{17}$

**2.3    Factors**

**Key Terms**

1. factorizations    2. composite number    3. prime factorization

4. prime number    5. factors

**Objective 1**

1. 1, 7    3. 1, 7, 49    5. 1, 2, 5, 10

7. 1, 5, 25    9. 1, 2, 3, 5, 6, 10, 15, 30

**Objective 2**

11. neither    13. composite    15. composite

17. prime    19. prime

**Objective 3**

21. $2^2 \cdot 3$    23. $2^2 \cdot 7$    25. $2^3 \cdot 3$

27. $2^2 \cdot 3^3$    29. $2 \cdot 3^2 \cdot 5^2$

## 2.4    Writing a Fraction in Lowest Terms

### Key Terms

1. lowest terms
2. common factor
3. equivalent fractions

### Objective 1

1. no
3. no
5. yes

7. $\dfrac{1}{3}$
9. $\dfrac{2}{9}$
11. $\dfrac{2}{7}$

13. $\dfrac{3}{22}$

### Objective 2

15. $\dfrac{3 \cdot 3 \cdot 7}{2 \cdot 2 \cdot 3 \cdot 7} = \dfrac{3}{4}$
17. $\dfrac{2 \cdot 2 \cdot 3 \cdot 3 \cdot 5}{2 \cdot 3 \cdot 5 \cdot 7} = \dfrac{6}{7}$
19. $\dfrac{2 \cdot 2 \cdot 3 \cdot 3}{2 \cdot 3 \cdot 3 \cdot 3} = \dfrac{2}{3}$

21. $\dfrac{3 \cdot 5 \cdot 5}{2 \cdot 2 \cdot 5 \cdot 5 \cdot 5} = \dfrac{3}{20}$

### Objective 3

23. equivalent
25. not equivalent
27. equivalent

29. not equivalent

## 2.5    Multiplying Fractions

### Key Terms

1.    common factor      2.    denominator      3.    multiplication shortcut

4.    numerator

### Objective 1

1.    $\dfrac{35}{54}$        3.    $\dfrac{55}{24}$        5.    $\dfrac{5}{12}$

7.    $\dfrac{1}{96}$

### Objective 2

9.    $\dfrac{5}{12}$        11.    $\dfrac{2}{3}$        13.    $\dfrac{1}{6}$

15.    $\dfrac{1}{7}$

### Objective 3

17.    $\dfrac{4}{5}$        19.    42        21.    9

23.    $1\dfrac{1}{2}$

### Objective 4

25.    $\dfrac{3}{8}$ m$^2$        27.    $\dfrac{15}{32}$ m$^2$        29.    $\dfrac{5}{9}$ yd$^2$

## 2.6    Applications of Multiplication

### Key Terms

1.  indicator words    2.  product

### Objective 1

1.  1680 paperbacks    3.  $1500      5.  133 employees

7.  $500              9.  $104       11.  320 muffins

13.  160 pages        15.  90 games    17.  1688 votes

19.  312 square feet    21.  6 gallons    23.  $18,000

25.  $3375            27.  $16,000     29.  $36,000

## 2.7    Dividing Fractions

### Key Terms

1.    reciprocal        2.    indicator words        3.    quotient

### Objective 1

1.    $\dfrac{4}{3}$        3.    3        5.    $\dfrac{1}{10}$

### Objective 2

7.    $2\dfrac{2}{15}$        9.    $\dfrac{1}{4}$        11.    6

13.    $\dfrac{11}{15}$        15.    $\dfrac{16}{25}$        17.    $\dfrac{2}{9}$

19.    $1\dfrac{53}{75}$

### Objective 3

21.    54 dresses        23.    24 Brownies        25.    32 guests

27.    24 patties        29.    16 tumblers

## 2.8    Multiplying and Dividing Mixed Numbers

### Key Terms

1. simplify       2. round       3. mixed number

### Objective 1

1. $15; 13\dfrac{1}{3}$       3. $4; 5$       5. $42; 40\dfrac{3}{8}$

7. $45; 51$       9. $2; 2\dfrac{1}{2}$

### Objective 2

11. $1\dfrac{1}{5}; 1\dfrac{1}{9}$       13. $1; 1\dfrac{1}{4}$       15. $1\dfrac{3}{4}; 1\dfrac{2}{3}$

17. $8; 11\dfrac{1}{4}$       19. $1\dfrac{4}{5}; 1\dfrac{3}{4}$

### Objective 3

21. 60 yards; 65 yards       23. 246 yards; 248 yards

25. 58 ounces; $50\dfrac{2}{5}$ ounces       27. $17\dfrac{1}{2}$ dresses; 16 dresses

29. 0 pounds; $\dfrac{7}{8}$ pound

# Chapter 3 ADDING AND SUBTRACTING FRACTIONS

## 3.1    Adding and Subtracting Like Fractions

### Key Terms

1.   unlike fractions    2.   like fractions

### Objective 1

1.   unlike        3.   like        5.   unlike

### Objective 2

7.   $\dfrac{3}{4}$        9.   $1\dfrac{1}{2}$        11.   $1\dfrac{1}{8}$

13.   $1\dfrac{2}{5}$        15.   $1\dfrac{1}{3}$        17.   $\dfrac{7}{11}$ of the debt

### Objective 3
### Objective 4

19.   $\dfrac{1}{5}$        21.   $\dfrac{2}{3}$        23.   $\dfrac{5}{14}$

25.   $\dfrac{1}{2}$        27.   $\dfrac{3}{7}$        29.   $\dfrac{1}{3}$ of the garden

## 3.2    Least Common Multiples

**Key Terms**

1.    least common multiple                    2.    LCM

**Objective 1**

1.    14                3.    84                5.    105

**Objective 2**

7.    80                9.    70

**Objective 3**

11.    336              13.    480              15.    420

**Objective 4**

17.    110              19.    108              21.    180

**Objective 5**

23.    18               25.    60               27.    105

29.    54

### 3.3 Adding and Subtracting Unlike Fractions

**Key Terms**

1. least common denominator

2. LCD

**Objective 1**

1. $\dfrac{5}{6}$

3. $\dfrac{23}{30}$

5. $\dfrac{43}{60}$

7. $\dfrac{7}{8}$

9. $\dfrac{19}{24}$ ton

**Objective 2**

11. $\dfrac{11}{15}$

13. $\dfrac{5}{6}$

15. $\dfrac{13}{21}$

17. $\dfrac{3}{4}$

19. $\dfrac{29}{54}$

**Objective 3**

21. $\dfrac{3}{8}$

23. $\dfrac{5}{36}$

25. $\dfrac{5}{8}$

27. $\dfrac{37}{50}$

29. $\dfrac{11}{24}$ of the goal

## 3.4 Adding and Subtracting Mixed Numbers

### Key Terms

1. regrouping when subtracting fractions

2. regrouping when adding fractions

### Objective 1

1. $10; 9\frac{4}{7}$

3. $36; 35\frac{17}{24}$

5. $14; 13\frac{32}{63}$

7. $4$ cans; $4\frac{5}{24}$ cans

9. $7$ hours; $6\frac{3}{8}$ hours

### Objective 2

11. $4; 4\frac{1}{2}$

13. $0; \frac{35}{48}$

15. $22; 22\frac{1}{4}$

17. $165; 164\frac{11}{24}$

19. $10$ hours; $8\frac{7}{8}$ hours

21. $6$ yd$^3$; $5\frac{1}{24}$ yd$^3$

23. $14; 15\frac{1}{20}$

### Objective 3

25. $8; 8\frac{1}{6}$

27. $5; 5\frac{7}{24}$

29. $2; 1\frac{5}{6}$

## 3.5    Order Relations and the Order of Operations

### Key Terms

1.  $<$                       2.  $>$

### Objective 1

1.  $<$            3.  $>$            5.  $>$

7.  $>$            9.  $>$

### Objective 2

11.  $\dfrac{1}{4}$        13.  $4\dfrac{17}{27}$        15.  $\dfrac{64}{121}$

17.  $2\dfrac{46}{49}$

### Objective 3

19.  $2\dfrac{2}{3}$        21.  $\dfrac{4}{25}$        23.  $1\dfrac{5}{12}$

25.  $\dfrac{1}{8}$        27.  $1\dfrac{1}{6}$        29.  $\dfrac{1}{4}$

# Chapter 4 DECIMALS

## 4.1   Reading and Writing Decimals
### Key Terms

1. decimals  2. place value  3. decimal point

### Objective 1

1. $\dfrac{8}{10}$; 0.8; eight tenths

3. $\dfrac{58}{100}$; 0.58; fifty-eight hundredths

### Objective 2

5. 2; 5  7. 3; 6

9. tens; ones; tenths; hundredths; thousandths

### Objective 3

11. seven thousandths

13. three and fourteen ten-thousandths

15. ten and eight hundred thirty five thousandths

17. ninety seven and eight thousandths  19. 11.009

21. 300.0023

23. seven and two hundred two thousandths

### Objective 4

25. $\dfrac{1}{1000}$  27. $20\dfrac{1}{2000}$  29. $\dfrac{19}{20}$

## 4.2    Rounding Decimals

### Key Terms

1. decimal places     2. rounding

### Objective 1

1. up

### Objective 2

| | | | | | |
|---|---|---|---|---|---|
| 3. | 17.9 | 5. | 785.498 | 7. | 54.40 |
| 9. | 989.990 | 11. | 283.05; 283.0 | 13. | 21.77; 21.8 |
| 15. | 1.44; 1.4 | 17. | 78.70; 78.7 | | |

### Objective 3

| | | | | | |
|---|---|---|---|---|---|
| 19. | $79 | 21. | $226 | 23. | $11,840 |
| 25. | $1.25 | 27. | $112.01 | 29. | $1028.67 |

## 4.3 Adding and Subtracting Decimals

**Key Terms**

1. front end rounding        2. estimating

**Objective 1**

1. 92.49      3. 105.43      5. 72.453

7. 48.35      9. 123.6802 in.

**Objective 2**

11. 115.8      13. 42.566      15. 58.32

17. 24.016 ft

**Objective 3**

19. 78; 82.91      21. 5; 4.838      23. 6; 5.53

25. 20 hr; 17.85 hr      27. $10; $8.71      29. 80,160 mi; 80,611.3 mi

## 4.4     Multiplying Decimals

### Key Terms

  **1.**    factor        2.    decimal places       3.    product

### Objective 1

  1.    0.2279           3.    90.71         5.    1.5548

  7.    0.0037           9.    \$163.08       11.    \$9.44

 13.    43.0             15.    0.430         17.    0.00430

 19.    348.04 sq ft     21.    \$2105.99

### Objective 2

 23.    300; 288.26      25.    240; 218.4756     27.    80; 43.548

 29.    3200; 3033.306

## 4.5    Dividing Decimals

**Key Terms**

1. dividend
2. repeating decimal
3. quotient
4. divisor

**Objective 1**

1. 1.794
3. 2.359
5. 16.589

**Objective 2**

7. 3.796
9. 53,950.943
11. 33.3 miles per gallon

**Objective 3**

13. reasonable
15. unreasonable
17. unreasonable
19. reasonable

**Objective 4**

21. 20.31
23. 54.02
25. 96.61
27. 7.91
29. 53.548

## 4.6　Writing Fractions as Decimals

### Key Terms

1. equivalent
2. numerator
3. denominator
4. mixed number

### Objective 1

1. 6.5
3. 2.667
5. 0.091
7. 0.6
9. 4.111
11. 0.15
13. 19.708

### Objective 2

15. $<$
17. $>$
19. $>$
21. $>$
23. $0.466, \frac{7}{15}, \frac{9}{19}$
25. $\frac{3}{11}, 0.29, \frac{1}{3}$
27. $\frac{11}{13}, 0.8462, \frac{6}{7}$
29. $0.01666, 0.1666, 0.16666, \frac{1}{6}$

## Chapter 5 RATIO AND PROPORTION

### 5.1    Ratios

### Key Terms

1.    ratio

2.    numerator; denominator

### Objective 1

1.    $\dfrac{3}{4}$

3.    $\dfrac{25}{19}$

5.    $\dfrac{17}{27}$

7.    $\dfrac{7}{3}$

### Objective 2

9.    $\dfrac{13}{4}$

11.    $\dfrac{6}{5}$

13.    $\dfrac{5}{6}$

15.    $\dfrac{3}{4}$

17.    $\dfrac{2}{11}$

19.    $\dfrac{8}{5}$

### Objective 3

21.    $\dfrac{2}{7}$

23.    $\dfrac{9}{5}$

25.    $\dfrac{5}{6}$

27.    $\dfrac{5}{8}$

29.    $\dfrac{8}{1}$

## 5.2 Rates

### Key Terms

1. unit rate
2. cost per unit
3. rate

### Objective 1

1. $\dfrac{3 \text{ miles}}{1 \text{ minute}}$

3. $\dfrac{7 \text{ dresses}}{1 \text{ woman}}$

5. $\dfrac{15 \text{ gallons}}{1 \text{ hour}}$

7. $\dfrac{7 \text{ pills}}{1 \text{ patient}}$

9. $\dfrac{32 \text{ pages}}{1 \text{ chapter}}$

### Objective 2

11. $15/hour

13. $110/day

15. $13.64/hour

17. $\dfrac{1}{2}$ crate/minute; 2 minutes/crate

19. $9.18/hour

21. $2.58/share

23. approximately 62 miles/hour

### Objective 3

25. 16 ounces for $0.89

27. 10 for $4.19

29. 5 cans for $2.75

## 5.3 Proportions

**Key Terms**

   **1.**   proportion      **2.**   cross products

**Objective 1**

1. $\dfrac{11}{15} = \dfrac{22}{30}$       3. $\dfrac{24}{30} = \dfrac{8}{10}$       5. $\dfrac{14}{21} = \dfrac{10}{15}$

7. $\dfrac{1\frac{1}{2}}{4} = \dfrac{21}{56}$       9. $\dfrac{6\frac{2}{5}}{12} = \dfrac{8}{3}$

**Objective 2**

11. $\dfrac{4}{3} = \dfrac{3}{4}$; false      13. $\dfrac{6}{5} = \dfrac{6}{5}$; true      15. $\dfrac{5}{3} = \dfrac{3}{4}$; false

17. $\dfrac{9}{5} = \dfrac{9}{5}$; true      19. $\dfrac{7}{2} = \dfrac{4}{1}$; false

**Objective 3**

21.   $270 = 270$; true      23.   $396 = 264$; false      25.   $165\frac{3}{5} = 165\frac{3}{5}$; true

27.   $114 = 342$; false      29.   $12.814 = 12.07$; false

## 5.4     Solving Proportions

### Key Terms

  **1.**   proportion        2.   cross products       3.   ratio

### Objective 1

  1.   9            3.   36         5.   21

  7.   5            9.   45       11.   77

13.   18         15.   33

### Objective 2

17.   1           19.   $2\frac{1}{2}$       21.   $3\frac{3}{8}$

23.   $1\frac{1}{2}$       25.   3         27.   8

29.   6

## 5.5 Solving Application Problems with Proportions

### Key Terms

1. ratio
2. rate

### Objective 1

1. $112.50
3. $15
5. 8 pounds
7. $818.40
9. 960 miles
11. $440
13. $4399.50
15. $122.50
17. $22.50
19. $2160
21. 480 minutes or 8 hours
23. $62\frac{1}{2}$ minutes
25. 76.8 feet
27. $2.56
29. approximately 11 days

## Chapter 6 PERCENTS

### 6.1    Basics of Percent

**Key Terms**

1.  ratio                    2.  percent                 3.  decimals

**Objective 1**

1.  43%                    3.  45%

**Objective 2**

5.  0.42                    7.  0.04                9.  0.025

11.  0.00256

**Objective 3**

13.  20%                    15.  56.4%              17.  550%

**Objective 4**

19.  $19                     21.  $228                23.  $1040

**Objective 5**

25.  125 signs               27.  24 copies           29.  4 homes

## 6.2   Percents and Fractions

### Key Terms

   1.   lowest terms     2.   percent

### Objective 1

   1.  $\dfrac{3}{25}$      3.  $\dfrac{5}{8}$      5.  $\dfrac{1}{6}$

   7.  $\dfrac{1}{200}$     9.  $1\dfrac{2}{5}$

### Objective 2

  11.  70%      13.  48%      15.  94%

  17.  380%     19.  740%

### Objective 3

  21.  0.5; 50%     23.  0.25; 25%     25.  $\dfrac{7}{8}$; 0.875

  27.  $\dfrac{1}{3}$; 0.333     29.  $\dfrac{13}{40}$; 32.5%

## 6.3 Using the Percent Proportion and Identifying the Components in a Percent Problem

**Key Terms**

1. whole          2. part          3. percent proportion

**Objective 1**

1. $\dfrac{\text{part}}{\text{whole}} = \dfrac{\text{percent}}{100}$

**Objective 2**

3. 800          5. 12          7. 87.5

9. 5%

**Objective 3**

11. 83%          13. 42%          15. 17%

**Objective 4**

17. 384          19. 78          21. unknown

**Objective 5**

23. 29.81          25. unknown          27. unknown; $\dfrac{x}{1500} = \dfrac{7}{100}$

29. unknown; $\dfrac{x}{40} = \dfrac{15}{100}$

## 6.4    Using Proportions to Solve Percent Problems

### Key Terms

**1.**  cross products      2.    percent proportion

### Objective 1

1.    280          3.    87.5          5.    24.5

7.    $84          9.    $210

### Objective 2

11.    300          13.    500          15.    3500

17.    800 applications    19.    150 students

### Objective 3

21.    0.05%          23.    5000%          25.    55%

27.    22%          29.    85%

## 6.5 Using the Percent Equation

### Key Terms

1. percent
2. percent equation

### Objective 1

1. 644
3. 106.4
5. 1.4

7. 14 clients
9. $100.50

### Objective 2

11. 160
13. 2160
15. 22.8

17. 640 gallons
19. 500 employees

### Objective 3

21. 20%
23. 5%
25. 244.4%

27. 25%
29. 23.3%

## 6.6    Solving Application Problems with Percent

### Key Terms

1.  commission
2.  percent of increase or decrease
3.  sales tax
4.  discount

### Objective 1

1.  $3.50; $53.50
3.  $6.03; $73.03
5.  5%
7.  $810

### Objective 2

9.  $155.63
11.  $3000
13.  4%
15.  $2196.72

### Objective 3

17.  $30; $170
19.  $10.28; $195.22
21.  $1408.64
23.  $36.18

### Objective 4

25.  25%
27.  131.25%
29.  11.4%

## 6.7    Simple Interest

### Key Terms

1. rate of interest
2. interest
3. interest formula
4. simple interest
5. principal

### Objective 1

1. $24
3. $2122
5. $32.40
7. $8
9. $36.90
11. $780
13. $1170
15. $120
17. $63.75

### Objective 2

19. $3075
21. $1125
23. $27,720
25. $32,548
27. $2149
29. $1230

### 6.8    Compound Interest

### Key Terms

1.    compound interest    2.    compound amount    3.    compounding

### Objective 1; Objective 2

1.    $2100          3.    $2315.25; $315.25

### Objective 3

5.    $4867.20_        7.    $4920          9.    $3595.52

### Objective 4

11.    $1262.50        13.    $11,277        15.    $60.47

17.    $19,815.73

### Objective 5

19.    $1272.30; $272.30          21.    $20,724.48; $7924.48

23.    $34,730.06; $13,330.06          25.    $111,212.40; $33,212.40

27.    $1302.30; $302.30          29.    $14,282.10; $5282.10

## Chapter 7 GEOMETRY

### 7.1    Lines and Angles

**Key Terms**

1.  ray
2.  perpendicular lines
3.  obtuse angle
4.  point
5.  angle
6.  line
7.  acute angle
8.  degrees
9.  parallel lines
10.  line segment
11.  intersecting lines
12.  right angle
13.  vertical angles
14.  congruent angles
15.  supplementary angles
16.  complementary angles
17.  alternate interior angles
18.  corresponding angles

### Objective 1

1.  line segment; $\overline{CD}$    3.    line, $\overleftrightarrow{EF}$

### Objective 2

5.    parallel

### Objective 3

7.    $\angle COD$ or $\angle DOC$    9.    $\angle MON$ or $\angle NOM$

### Objective 4

11.    obtuse    13.    straight

### Objective 5

15.    perpendicular    17.    neither

### Objective 6

19.    $\angle BAC, \angle CAD$;  $\angle FAE, \angle EAD$

21.    $\angle MKN, \angle NKO$;  $\angle NKO, \angle LKO$;  $\angle LKO, \angle LKM$;  $\angle LKM, \angle MKN$

23.    complement: 74°; supplement: 164°

25.    $\angle POQ \cong \angle NOM$;  $\angle NOP \cong \angle MOQ$; $m\angle NOM = 34°$;  $m\angle NOP = m\angle MOQ = 146°$

27.    corresponding angles: $\angle 1$ and $\angle 5$; $\angle 2$ and $\angle 6$; $\angle 3$ and $\angle 7$; $\angle 4$ and $\angle 8$
       alternate interior angles: $\angle 2$ and $\angle 7$; $\angle 4$ and $\angle 5$
       $m\angle 1 = 100°$; $m\angle 2 = 80°$; $m\angle 3 = 80°$; $m\angle 4 = 100°$
       $m\angle 5 = 100°$; $m\angle 6 = 80°$; $m\angle 7 = 80°$; $m\angle 8 = 100°$

29.    corresponding angles: $\angle 1$ and $\angle 5$; $\angle 2$ and $\angle 6$; $\angle 3$ and $\angle 7$; $\angle 4$ and $\angle 8$
       alternate interior angles: $\angle 4$ and $\angle 6$; $\angle 3$ and $\angle 5$
       $m\angle 1 = 37°$; $m\angle 2 = 143°$; $m\angle 3 = 37°$; $m\angle 4 = 143°$
       $m\angle 5 = 37°$; $m\angle 6 = 143°$; $m\angle 7 = 37°$; $m\angle 8 = 143°$

## 7.2    Rectangles and Squares

### Key Terms

1. area
2. rectangle
3. perimeter
4. square

### Objective 1

1. $P = 24$ cm; $A = 32$ cm$^2$

3. $P = 36$ cm; $A = 17$ cm$^2$

5. $P = 22$ yd; $A = 29\frac{1}{4}$ yd$^2$

7. $P = 281.2$ cm; $A = 3859.68$ cm$^2$

9. $A = 7248$ ft$^2$

### Objective 2

11. $P = 36$ m; $A = 81$ m$^2$

13. $P = 31.2$ ft; $A = 60.84$ ft$^2$

15. $P = 5\frac{3}{5}$ in.; $A = 1\frac{24}{25}$ in.$^2$

17. $P = 12.4$ cm; $A = 9.61$ cm$^2$

19. $P = 18\frac{2}{3}$ mi; $A = 21\frac{7}{9}$ mi$^2$

### Objective 3

21. $P = 30$ ft; $A = 18$ ft$^2$

23. $P = 42$ yd; $A = 50$ yd$^2$

25. $P = 24$ cm; $A = 31$ cm$^2$

27. $P = 42$ yd; $A = 54$ yd$^2$

29. $P = 56$ cm; $A = 171$ cm$^2$

Answers to Worksheets for Classroom or Lab Practice

## 7.3    Parallelograms and Trapezoids

**Key Terms**

1. parallelogram
2. trapezoid
3. perimeter
4. area

**Objective 1**

1. 168 m
3. 30 ft
5. 22 cm
7. 27 in.$^2$
9. 713 yd$^2$
11. $11\frac{1}{4}$ m$^2$
13. 310 ft$^2$
15. $780

**Objective 2**

17. 708.8 cm
19. 1106 m$^2$
21. 60 in.$^2$
23. $9\frac{5}{8}$ in.$^2$
25. 4190 cm$^2$
27. 15,504 ft$^2$
29. $700

## 7.4   Triangles

### Key Terms

1. triangle
2. base
3. height

### Objective 1

1. 25 yd
3. $24\frac{1}{2}$ ft
5. 37.2 ft
7. $10\frac{3}{4}$ ft
9. 17.7 m

### Objective 2

11. 1260 m$^2$
13. $21\frac{3}{4}$ ft$^2$
15. 15.81 m$^2$
17. 510 m$^2$
19. 534 m$^2$
21. 1940 yd$^2$

### Objective 3

23. 17°
25. 81°
27. 51°
29. 45°

## 7.5    Circles

### Key Terms

1. radius
2. circumference
3. circle
4. $\pi$ (pi)
5. diameter

### Objective 1

1. 86 m
3. 13.25 m
5. 4 ft
7. $6\frac{1}{4}$ yd

### Objective 2

9. 94.2 m
11. 28.3 yd
13. 188.4 cm

### Objective 3

15. 43.0 m$^2$
17. 22.3 yd$^2$
19. 57 cm$^2$
21. 2101.3 m$^2$
23. 18.2 ft$^2$
25. $320.87

### Objective 4  *Other answers are possible.*

27. polynomial; polyglot
29. octagon; octopus

## 7.6    Volume and Surface Area

### Key Terms

1. rectangular solid    2. cylinder    3. sphere

4. volume    5. pyramid    6. cone

7. surface area

### Objective 1

1. $2744 \text{ in.}^3$    3. $95.2 \text{ cm}^3$    5. $2310 \text{ m}^3$

### Objective 2

7. $0.9 \text{ in.}^3$    9. $3267.5 \text{ ft}^3$

### Objective 3

11. $471 \text{ ft}^3$    13. $490.6 \text{ m}^3$    15. $1105.3 \text{ in.}^3$

### Objective 4

17. $150.7 \text{ yd}^3$    19. $22{,}344 \text{ m}^3$

### Objective 5

21. $558 \text{ in.}^2$    23. $2670 \text{ mm}^2$    25. $130 \text{ cm}^2$

### Objective 6

27. $183.6 \text{ ft}^2$    29. $138.2 \text{ ft}^2$

## 7.7    Pythagorean Theorem

### Key Terms

1. right triangle    2. hypotenuse    3. legs

### Objective 1

1. 4.123    3. 1.414    5. 8.660

7. 12.042

### Objective 2

9. 9.8 in.    11. 12.1 in.    13. 1.0 in.

15. 8.5 in.    17. 13 ft

### Objective 3

19. 9.5 ft    21. 5 ft    23. 30 ft

25. 8 ft    27. 7.2 ft    29. 19.1 ft

### 7.8    Congruent and Similar Triangles

### Key Terms

1. congruent triangles  2. congruent figures    3. similar triangles

4. similar figures

## Objective 1

1.    $\angle 1$ and $\angle 4$; $\angle 2$ and $\angle 5$; $\angle 3$ and $\angle 6$; $\overline{AB}$ and $\overline{DE}$; $\overline{BC}$ and $\overline{EF}$; $\overline{AC}$ and $\overline{DF}$

## Objective 2

3.  ASA                 5.  SSS                 7.  SSS

9.  ASA

## Objective 3

11.    $\angle P$ and $\angle S$; $\angle M$ and $\angle Q$; $\angle N$ and $\angle R$; $\overline{MN}$ and $\overline{QR}$; $\overline{PN}$ and $\overline{SR}$; $\overline{PM}$ and $\overline{SQ}$

13.    $\angle G$ and $\angle T$; $\angle H$ and $\angle R$; $\angle K$ and $\angle S$; $\overline{GH}$ and $\overline{TR}$; $\overline{GK}$ and $\overline{TS}$; $\overline{HK}$ and $\overline{RS}$

15.    $\dfrac{5}{3}, \dfrac{5}{3}, \dfrac{5}{3}$          17.  $\dfrac{5}{12}, \dfrac{5}{12}, \dfrac{5}{12}$

## Objective 4

19.    $a = 15, b = 9$        21.    $x = 12.75; y = 15$

23.    $ABC$ : 54 in.; $DEF$ : 67.5 in.

## Objective 5

25.  39 m                27.  36 m                29.  144 m

# Chapter 8 STATISTICS

## 8.1 Circle Graphs

### Key Terms

1. circle graph
2. protractor

### Objective 1

1. $10,400

3. $400

5. $\dfrac{300}{1400}$ or $\dfrac{3}{14}$

### Objective 2

7. business

9. $\dfrac{1800}{11,600}$ or $\dfrac{9}{58}$

11. $\dfrac{600}{2400}$ or $\dfrac{1}{4}$

13. $285,000

15. $95,000

17. $142,500

19. $\dfrac{142,500}{95,000}$ or $\dfrac{3}{2}$

21. $\dfrac{285,000}{237,500} = \dfrac{6}{5}$

23. computer science

25. $\dfrac{18\%}{6\%}$ or $\dfrac{576}{192}$ or $\dfrac{3}{1}$

27. 29%

### Objective 3

29. (a) mysteries: 108°; biographies: 54°; cookbooks: 36°; romance novels: 90°; science: 54°; business: 18°

(b)

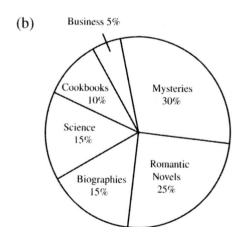

682

## 8.2    Bar Graphs and Line Graphs

### Key Terms

1. line graph
2. double-bar graph
3. bar graph
4. comparison line graph

### Objective 1

1. 1000 students
3. 1800 students
5. 2006
7. 400 students

### Objective 2

9. 350 female freshman

11. $\dfrac{500}{350}$ or $\dfrac{10}{7}$

13. $\dfrac{850}{550}$ or $\dfrac{17}{11}$

### Objective 3

15. September
17. $20
19. $\dfrac{60}{20}$ or $\dfrac{3}{1}$
21. $30

### Objective 4

23. $1,000,000
25. $2,500,00000
27. $3,000,000
29. 2004

## 8.3　Frequency Distributions and Histograms

**Key Terms**

1. histogram
2. frequency distribution

**Objective 1**

1. ‖ ; 2

3. 卌 | ; 6

5. ‖‖ ; 4

**Objective 2**

7. 卌 ; 5

9. ‖‖ ; 3

11. ‖‖ ; 3

13. 110–129

15. | ; 1

17. 卌 ‖ ; 7

19. | ; 1

21. 80–89 and 120–129

23.

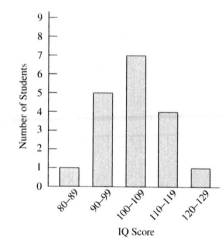

25. 16–20

27. 120 members

29. 30 members

## 8.4    Mean, Median, and Mode

### Key Terms

1. dispersion
2. mode
3. weighted mean
4. mean
5. median
6. bimodal
7. range

### Objective 1

1. 60.3
3. 64.5
5. 52.7
7. 5.4

### Objective 2

9. 18.6
11. 4.7
13. 3.1
15. 2.5

### Objective 3

17. 232
19. 632
21. 25
23. 239.5

### Objective 4

25. 3
27. 24, 35, 39
29. 8

# Chapter 9 THE REAL NUMBER SYSTEM

## 9.1 Exponents, Order of Operations, and Inequality

### Key Terms

1. exponential expression      2. base

3. exponent

### Objective 1

1. 27      3. $\dfrac{16}{81}$      5. 13.824

### Objective 2

7. 45      9. $\dfrac{29}{16}$

### Objective 3

11. $-8$      13. 48      15. 96

### Objective 4

17. false      19. false

### Objective 5

21. $7 = 13 - 6$      23. $30 - 7 > 20$      25. $20 \geq 2 \cdot 7$

### Objective 6

27. $\dfrac{2}{3} < \dfrac{3}{4}$      29. $.0002 < .002$

**9.2     Variables, Expressions, and Equations**

**Key Terms**

1.  equation          2.  variable          3.  algebraic expression

4.  solution

**Objective 1**

1.  8          3.  $-\dfrac{9}{2}$          5.  $\dfrac{28}{13}$

**Objective 2**

7.  $1 + 3x$          9.  $10x + 21$          11.  $8x - 11$

**Objective 3**

13.  no          15.  no          17.  yes

**Objective 4**

19.  $5x + 2 = 23$          21.  $6(5 + x) = 19$          23.  $61 - 7x = 13 + x$

**Objective 5**

25.  expression          27.  equation          29.  expression

## 9.3    Real Numbers and the Number Line

### Key Terms

1. whole numbers
2. opposite
3. integers
4. natural numbers
5. absolute value
6. number line
7. irrational number
8. coordinate
9. negative number
10. positive number
11. real numbers
12. set-builder notation
13. rational number

### Objective 1

1. −75 pounds
3. −396 meters

5.
7.

### Objective 2

9. −6.01
11. −4
13. true
15. false

### Objective 3

17. 25
19. −22
21. 0

23. $\dfrac{5}{7}$

### Objective 4

25. −10
27. $\dfrac{5}{6}$
29. 2

## 9.4 Adding Real Numbers

### Key Terms

1. sum      2. addends

### Objective 1

1. $-40$      3. $-18$      5. $\dfrac{7}{5}$

### Objective 2

7. $-5$      9. $0$      11. $\dfrac{1}{35}$

### Objective 3

13. true      15. false      17. false

### Objective 4

19. $-16$      21. $-6$      23. $-\dfrac{7}{8}$

### Objective 5

25. $-8+(-4)+(-11); -23$      27. $-10+[20+(-4)]; 6$

29. $495

## 9.5    Subtracting Real Numbers

### Key Terms

1. minuend        2. subtrahend        3. difference

### Objective 1

1. 3        3. 0        5. −7

### Objective 2

7. 7        9. 0        11. 4.4

13. $-\dfrac{1}{30}$

### Objective 3

15. 18        17. 18        19. −2

21. $-\dfrac{23}{18}$ or $-1\dfrac{5}{8}$

### Objective 4

23. $-4-4;\ -8$        25. $(-4+12)-9;\ -1$        27. −51.2°C

29. 37°F

**9.6    Multiplying and Dividing Real Numbers**

**Key Terms**

1.    quotient        2.    reciprocals        3.    product

**Objective 1**

1.    −28        3.    $-\dfrac{7}{12}$

**Objective 2**

5.    40        7.    2.73

**Objective 3**

9.    $-\dfrac{1}{6}$        11.    0        13.    −2.5

**Objective 4**

15.    70        17.    $\dfrac{16}{21}$

**Objective 5**

19.    7        21.    17

**Objective 6**

23.    $(-7)(3)+(-7); \ -28$        25.    $-12+\dfrac{49}{-7}; \ -19$

**Objective 7**

27.    $\dfrac{2}{3}x=-7$        29.    $\dfrac{x}{-4}=1$

## 9.7    Properties of Real Numbers

**Key Terms**

1. identity element for addition

2. identity element for multiplication

**Objective 1**

1. 4

3. $[10 + (-9)]$

5. $\left( \frac{1}{4} \cdot 2 \right)$

**Objective 2**

7. (4)

9. $4a$

11. $[(-r)(-p)]$

**Objective 3**

13. 4

15. 12

17. $\dfrac{30}{30}$ or 1

**Objective 4**

19. 4; inverse

21. 0; identity

23. $-\dfrac{6}{17}$ ; inverse

**Objective 5**

25. $4b + 8$

27. $-10y + 18z$

29. $-14(x + y)$

## 9.8   Simplifying Expressions

**Key Terms**

1.   numerical coefficient

2.   term

3.   like terms

**Objective 1**

1.   $8x + 27$

3.   $5 + s$

5.   $35n - 9$

**Objective 2**

7.   $-2$

9.   $1$

11.   $\dfrac{7}{9}$

**Objective 3**

13.   like

15.   unlike

17.   unlike

**Objective 4**

19.   $5r - 4$

21.   $1.7y^2 - .5xy$

23.   $-1.5y + 16$

**Objective 5**

25.   $6x + 12 + 4x = 10x + 12$

27.   $3(9 + 2x) + 4x = 10x + 27$

29.   $4(2x - 6x) + 6(x + 9) = -10x + 54$

# Chapter 10 EQUATIONS, INEQUALITIES, AND APPLICATIONS

## 10.1   The Addition Property of Equality

### Key Terms

1.   equivalent equations
2.   linear equation
3.   solution set

### Objective 1

1.   yes
3.   no
5.   no

### Objective 2

7.   20
9.   −5
11.   $\dfrac{3}{2}$

13.   −5
15.   −12.8
17.   −10

### Objective 3

19.   7
21.   −8
23.   $\dfrac{5}{4}$

25.   0
27.   $\dfrac{1}{3}$
29.   7.2

**10.2   The Multiplication Property of Equality**

**Key Terms**

1.  multiplication property of equality

2.  addition property of equality

**Objective 1**

| | | | | | |
|---|---|---|---|---|---|
| 1. | 3 | 3. | 3 | 5. | $-14$ |
| 7. | $\dfrac{8}{7}$ | 9. | $\dfrac{7}{9}$ | 11. | 3.6 |
| 13. | $-8.2$ | 15. | $-2.7$ | | |

**Objective 2**

| | | | | | |
|---|---|---|---|---|---|
| 17. | 25 | 19. | 8 | 21. | $-8$ |
| 23. | $-1.5$ | 25. | 10 | 27. | $-24$ |
| 29. | $-7$ | | | | |

## 10.3    More on Solving Linear Equations

**Key Terms**

1. contradiction    2. conditional equation    3. identity

**Objective 1**

1. $\dfrac{5}{2}$    3. 2    5. $-\dfrac{1}{4}$

7. $-\dfrac{1}{5}$

**Objective 2**

9. 2    11. $\dfrac{53}{11}$    13. 30

15. $-3$

**Objective 3**

17. none    19. none    21. none

23. infinitely many

**Objective 4**

25. $\dfrac{17}{p}$    27. $4x$    29. $x + 28$

## 10.4 An Introduction to Applications of Linear Equations

### Key Terms

1. supplementary angles

2. complementary angles

3. right angle    4. straight angle    5. consecutive integers

### Objective 1

1. Read the problem; assign a variable to represent the unknown; write an equation; solve the equation; state the answer; check the answer.

### Objective 2

3. $4x - 2 = 3 + 6x; \; -\dfrac{5}{2}$

5. $6(x - 4) = -2x; \; 3$

7. $4x + 7 = 6x - 5; \; 6$

### Objective 3

9. $x + (x + 21) = 439; \; 209$ votes

11. $x + (x + 5910) = 34{,}730;$ Mt. Rainier: 14,410 ft; Mt. McKinley: 20,320 feet

13. $x + 2x + 8x = 176;$ cranberry juice: 16 oz; orange juice: 32 oz; ginger ale: 128 oz

15. $2x + 36 = 52; \; 8$ feet

### Objective 4

17. $20°$    19. $49°$    21. $55°$

### Objective 5

23. 76, 78    25. 27, 28    27. 13, 15

29. 75, 76, 77

## 10.5 Formulas and Additional Applications from Geometry

### Key Terms

1.  vertical angles
2.  formula
3.  perimeter
4.  area

### Objective 1

1.  $V = 24$
3.  $a = 36$
5.  $C = 40$
7.  $V = 100.48$

### Objective 2

9.  8 in.
11. 1.5 years
13. 3052.08 cm$^2$
15. 31,400 sq ft

### Objective 3

17. 35°, 35°
19. 150°, 150°
21. 54°, 126°
23. 148°, 148°

### Objective 4

25. $r = -\dfrac{a-s}{s}$ or $\dfrac{s-a}{s}$
27. $A = \dfrac{P}{1-rt}$
29. $n = \dfrac{S}{180} + 2$ or $n = \dfrac{S+360}{180}$

### 10.6    Solving Linear Inequalities

### Key Terms

1.    three-part inequality

2.    interval

3.    linear inequality    4.    inequalities

5.    interval notation

### Objective 1

1.    $(3, \infty)$;

2.    $(-1, 3)$;

5.    $(-3, 2]$;

### Objective 2

7.    $[1, \infty)$;

9.    $(-\infty, 0]$;

### Objective 3

11.    $(-2, \infty)$;

13.    $[0, \infty)$;

15.    $(-\infty, 4)$

### Objective 4

17.    $(-\infty, 19]$;

19.    $\left[ \dfrac{13}{12}, \infty \right)$;

### Objective 5

21.    89

23.    55 students

25.    all numbers greater than 5

Answers to Worksheets for Classroom or Lab Practice

## Objective 6

27. $[-5, -3)$;

29. $[-6, 3)$;

# Chapter 11 GRAPHS OF LINEAR EQUATIONS AND INEQUALITIES IN TWO VARIABLES

## 11.1    Reading Graphs; Linear Equations in Two Variables

### Key Terms

1.    line graph

2.    linear equation in two variables

3.    bar graph

4.    coordinates

5.    $x$-axis

6.    $y$-axis

7.    ordered pair

8.    rectangular (Cartesian) coordinate system

9.    quadrants

10.    origin

11.    plot

12.    scatter diagram

13.    table of values

14.    plane

### Objective 1

1.    2002, 2003, 2005

3.    2003, 2004

5.    325 M.B.A. degrees

### Objective 2

7.    $\left(\frac{1}{3}, -9\right)$

9.    (.2, .3)

### Objective 3

11.    not a solution

13.    a solution

15.    a solution

### Objective 4

17.(a) (−4, −5); (b) (2, 7); (c) $\left(-\frac{3}{2}, 0\right)$; (d) (−2, −1); (e) (−5, −7)

19.(a) (2, 4); (b) (0, 4); (c) (4, 4); (d) (−4, 4); (e) (.75, 4)

### Objective 5

21.    $\left(0, \frac{3}{2}\right), (−2, 0), (2, 3)$

23.    $(1, −3), (1, 0), (1, 5)$

25.    $(0, 4), (3, 0), \left(\frac{15}{4}, −1\right)$

Answers to Worksheets for Classroom or Lab Practice

## Objective 6

27.–29.

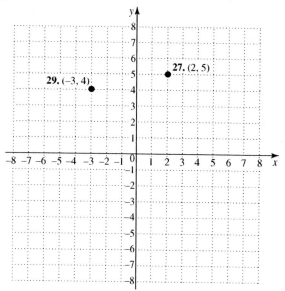

### 11.2    Graphing Linear Equations in Two Variables

### Key Terms

1.  *y*-intercept    2.  *x*-intercept    3.  graphing

4.  graph

### Objective 1

1.  (0, 3), (3, 0), (2, 1)

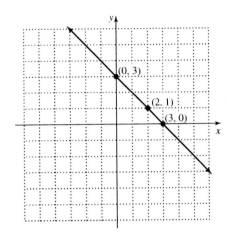

3.  (0, −4), (4, 0), (−2, −6)

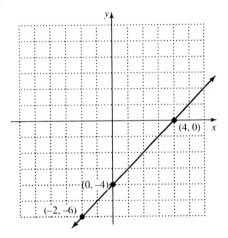

5.  (0, 2), (3, 0), (−3, 4)

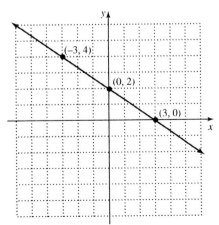

Answers to Worksheets for Classroom or Lab Practice

## Objective 2

7.

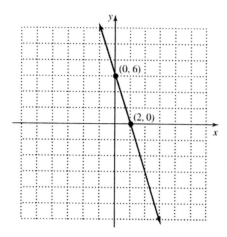

9.

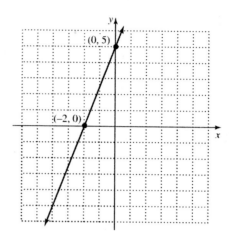

11.

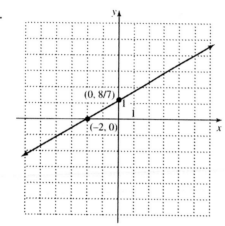

## Objective 3

13. $-3x - 2y = 0$

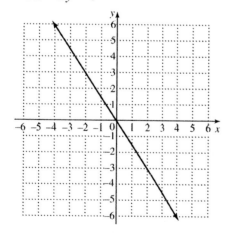

15. $x + 5y = 0$

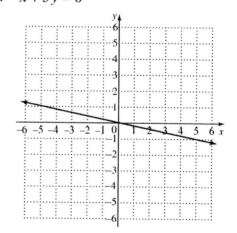

17. $4x = 3y$

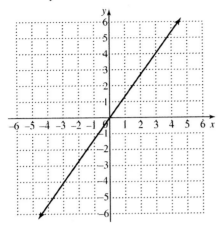

**Objective 4**

19. $y = -2$

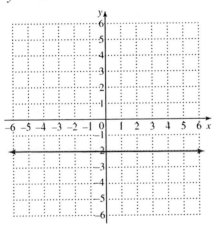

21. $x - 1 = 0$

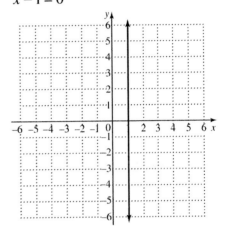

23. $y + 3 = 0$

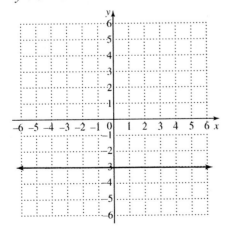

Answers to Worksheets for Classroom or Lab Practice

## Objective 5

25. (2000, 2435), (2001, 2350), (2002, 2265), (2003, 2180), (2004, 2095), (2005, 2010)

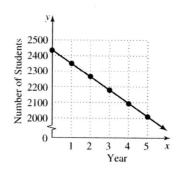

27. (2003, 325), (2004, 367), (2005, 409), (2006, 451)

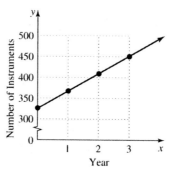

29. (1995, 2910), (2000, 2560), (2005, 2210), (2015, 1510)

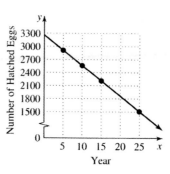

## 11.3   Slope of a Line

### Key Terms

1. perpendicular lines   2. slope        3. rise

4. parallel lines        5. run

### Objective 1

1. $-2$                  3. $-\dfrac{1}{14}$        5. $\dfrac{5}{3}$

7. $4$                   9. $-\dfrac{2}{5}$

### Objective 2

11. $\dfrac{1}{2}$       13. $\dfrac{4}{7}$         15. $0$

17. $-\dfrac{2}{5}$      19. $-\dfrac{2}{7}$

### Objective 3

21. $-5$; $5$; neither   23. $1$; $1$; parallel     25. $-2$; $-\dfrac{1}{4}$; neither

27. $-2$; $-\dfrac{5}{3}$; neither   29. $-4$; $-1$; neither

## 11.4    Equations of Lines

### Key Terms

1.  point-slope form    2.  standard form    3.  slope-intercept form

### Objective 1

1.  $y = \dfrac{3}{2}x - \dfrac{2}{3}$    3.  $y = -4x$    5.  $y = -3x + 3$

### Objective 2

7.

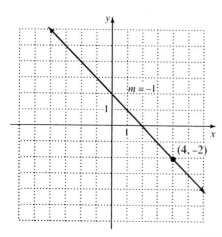

9.

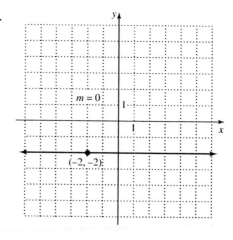

11.

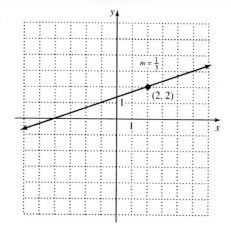

### Objective 3

13.  $x = -4$    15.  $y = \dfrac{4}{3}x - \dfrac{5}{3}$    17.  $y = \dfrac{2}{3}x + \dfrac{8}{3}$

19.  $y = -\dfrac{3}{2}x + 5$

## Objective 4

21.    $3x - 2y = 0$       23.    $2x + 3y = 3$       25.    $11x + y = 29$

## Objective 5

27.   (a)   (0, 686), (2, 620), (3, 592), (4, 570), (5, 542)

      (c)   $y = -\dfrac{144}{5}x + 686$

      (d)   \$426.80 (The year 2010 corresponds to $x = 9$)

      (b)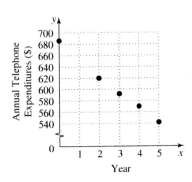

29.   (a)   (0, 10.1), (5, 16.2), (10, 26.0), (15. 29.1), (20, 32.5)

      (c)   $y = \dfrac{28}{25}x + 10.1$

      (d)   43.7% (The year 2015 corresponds to $x = 30$)

      (b)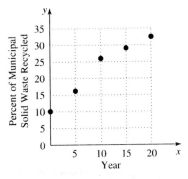

Answers to Worksheets for Classroom or Lab Practice

## 11.5   Graphing Linear Inequalities in Two Variables

### Key Terms

1.   boundary line   2.   linear inequality in two variables

### Objective 1

1.

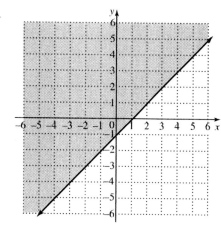

3.

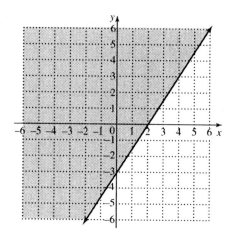

5.

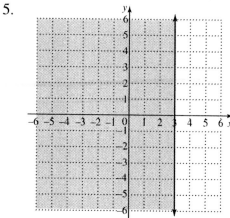

7.

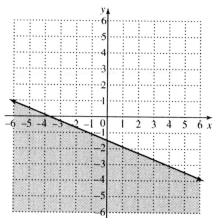

9.

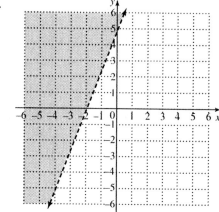

11.

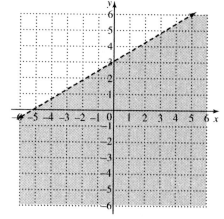

13.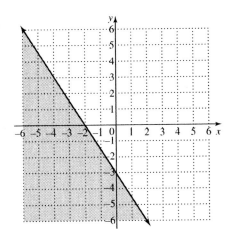

15.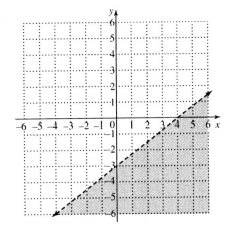

## Objective 2

17.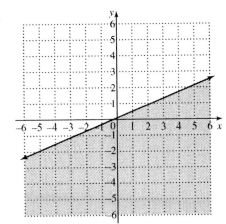

19.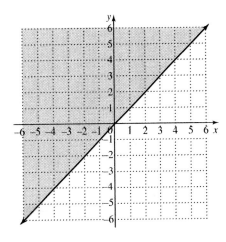

21.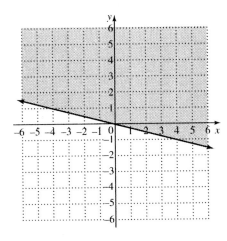

23.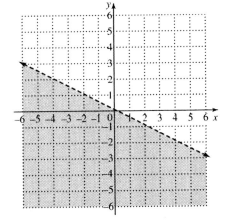

25.

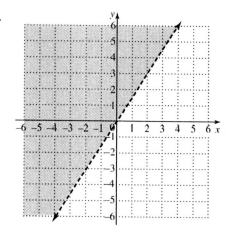

27.

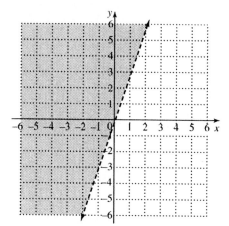

29.

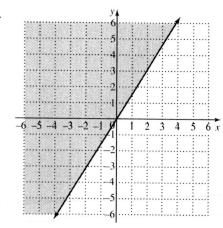

# Chapter 12 EXPONENTS AND POLYNOMIALS

## 12.1 Adding and Subtracting Polynomials
### Key Terms

1. degree of a term
2. descending powers
3. term
4. trinomial
5. polynomial
6. numerical coefficient
7. monomial
8. degree of a polynomial
9. binomial
10. like terms

### Objective 1

1. $-3z^3$

3. $-\frac{1}{4}r^3$

5. $8c^3 - 8c^2 - 6c + 6$

### Objective 2

7. $\frac{1}{2}x^2 - \frac{1}{2}x$ ; degree 2; binomial

9. $-5.7d^8 - 1.1d^5 + 3.2d^3 - d^2$ ; degree 8; none of these

### Objective 3

11.a. $-15$; b. 115

13.a. 71; b. $-19$

15. a. $-28$; b. 352

### Objective 4

17. $5m^3 - 2m^2 - 4m + 4$

19. $3r^3 + 7r^2 - 5r - 2$

### Objective 5

21. $-8w^3 + 21w^2 - 15$

23. $8b^4 - 4b^3 - 2b^2 - b - 2$

25. $7d^4 - 3d^2 - 19d + 12$

### Objective 6

27. $-8m^2n - 8m - 4n$  29. $-7x^2y + 3xy + 5xy^2$

## 12.2    The Product Rule and Power Rules for Exponents

### Key Terms

1.   power

2.   exponential expression   3.   base

### Objective 1

1.   $\dfrac{1}{243}$

3.   256; base: −4; exponent: 4

### Objective 2

5.   $7^7$

7.   $8c^{15}$

### Objective 3

9.   $7^{12}$

11.   $\left(\dfrac{1}{3}\right)^{15}$

### Objective 4

13.   $\dfrac{1}{9}x^8$

15.   $-.008a^{12}b^3$

### Objective 5

17.   $-\dfrac{8x^3}{125}$

19.   $-\dfrac{128a^7}{b^{14}}$

### Objective 6

21.   $x^{26}$

23.   $5^{11}x^{18}y^{37}$

25.   $7^3a^{10}b^{18}c^{19}$

### Objective 7

27.   $36x^5$

29.   $3p^8q^6$

**12.3    Multiplying Polynomials**

**Key Terms**

1. inner product
2. FOIL
3. outer product

**Objective 1**

1. $16y^7$

3. $6m + 14m^3 + 6m^4$

5. $-6z^6 - 18z^4 - 24z^2 - 12z$

7. $-35b^4 + 7b^2 - 28b^3$

9. $32m^3n + 16m^2n^2 + 56mn^3$

**Objective 2**

11. $6x^2 + 23x + 20$   13. $x^3 + 27$

15. $2r^3 + 3r^2 - 4r + 15$

17. $6m^5 + 4m^4 - 5m^3 + 2m^2 - 4m$

19. $y^4 - 4y^3 - 2y^2 + 12y + 9$

**Objective 3**

21. $6x^2 - 5xy - 6y^2$   23. $3 + 10a + 8a^2$

25. $-6m^2 - mn + 12n^2$

27. $x^2 - .2x - .15$   29. $x^2 - \dfrac{5x}{3} + \dfrac{4}{9}$

### 12.4   Special Products

### Key Terms

1. binomial     2. conjugate

### Objective 1

1. $25y^2 - 30y + 9$     3. $49 + 14x + x^2$     5. $4p^2 + 12pq + 9q^2$

7. $16y^2 - 5.6y + .49$     9. $9x^2 + 2xy + \dfrac{y^2}{9}$

### Objective 2

11. $144 - x^2$     13. $49x^2 - 9y^2$     15. $x^2 - .04$

17. $49m^2 - \dfrac{9}{16}$     19. $16x^2 - \dfrac{49}{36}$

### Objective 3

21. $a^3 - 9a^2 + 27a - 27$

23. $64x^3 + 48x^2 y + 12xy^2 + y^3$

25. $\dfrac{1}{8}t^3 + \dfrac{3}{2}t^2 u + 6tu^2 + 8u^3$

27. $x^4 + 8x^3 y + 24x^2 y^2 + 32xy^3 + 16y^4$

29. $16x^4 - 32x^3 y + 24x^2 y^2 - 8xy^3 + y^4$

## 12.5 Integer Exponents and the Quotient Rule

### Key Terms

1. power rule for exponents

2. base; exponent

3. product rule for exponents

### Objective 1

1. $-1$

3. $2$

5. $-2$

7. $0$

### Objective 2

9. $2r^7$

11. $-\dfrac{1}{64}$

13. $-\dfrac{3}{x^2}$

15. $4$

### Objective 3

17. $\dfrac{1}{2x^6}$

19. $\dfrac{x^6}{12^3 y^2}$ or $\dfrac{x^6}{1728 y^2}$

21. $\dfrac{p^8}{3^5 m^3}$ or $\dfrac{p^8}{243 m^3}$

### Objective 4

23. $\dfrac{9}{16 y^2}$

25. $\dfrac{1}{9xy}$

27. $a^{16} b^{22}$

29. $\dfrac{2^4 \cdot 3^8 x^{20}}{y^8}$ or $\dfrac{104,976 x^{20}}{y^8}$

## 12.6 Dividing a Polynomial by a Monomial

### Key Terms

1. dividend
2. quotient
3. divisor

### Objective 1

1. $2a^3 - 3a$

3. $5a^2 - \dfrac{9}{4}$

5. $1 + 3p^3$

7. $2 + 3y^3 - 7y^5$

9. $8p^2 - 7p - \dfrac{3}{p}$

11. $-\dfrac{r}{3s} - \dfrac{2}{3} + \dfrac{s}{r}$

13. $-4a^3 + \dfrac{3a^2 b}{4} - 3ab^2$

15. $-4p^2 - 3 - \dfrac{5}{p} + \dfrac{1}{7p^2}$

17. $2z^4 + 9z^2 - 4 + \dfrac{10}{3z}$

19. $\dfrac{m}{2} + \dfrac{7}{2} - \dfrac{21}{m}$

21. $2y^6 + 8y^3 - 41 - \dfrac{12}{y^3}$

23. $12 + 16x^3 + \dfrac{x^7}{2}$

25. $-3 + \dfrac{2}{y} - \dfrac{6}{y^2}$

27. $10d^2 + 4d - 7 - \dfrac{4}{d^2}$

29. $4y^2 + 5y - \dfrac{2}{3y}$

## 12.7 Dividing a Polynomial by a Polynomial

### Key Terms

1. divisor
2. quotient
3. dividend

### Objective 1

1. $6a - 5$
3. $p + 8$
5. $9a - 1$

7. $a - 7 + \dfrac{37}{2a + 3}$
9. $3m - 2 + \dfrac{8}{3m - 4}$
11. $5b - 3 + \dfrac{24}{b + 7}$

13. $3y^2 - 5y + 6$
15. $2m^2 + m - 5 + \dfrac{26}{3m + 2}$
17. $3x - 2 + \dfrac{34x - 35}{x^2 - 3x - 5}$

19. $3y^3 - 2y^2 + 2y - 1 + \dfrac{6y - 8}{4y^2 - 3}$
21. $y^2 - y + 1$

23. $3x^2 + 2x + 1$

25. $16x^4 + 24x^3 + 36x^2 + 54x + 81$

### Objective 2

27. $2r^2 - r + 5$ square units
29. $4y^2 + 24y + 100$ units

## 12.8  An Application of Exponents: Scientific Notation

### Key Terms

1. scientific notation     2. power rule          3. quotient rule

### Objective 1

1. $3.25 \times 10^2$       3. $2.3651 \times 10^4$     5. $9.54 \times 10^6$

7. $5.03 \times 10^{-2}$    9. $-2.2208 \times 10^{-4}$

### Objective 2

11. 72,000,000      13. 23,000       15. 0.0064

17. 0.04007         19. $-4.02$

### Objective 3

21. $2.53 \times 10^2$      23. $2 \times 10^2$        25. $2.1 \times 10^0$ or 2.1

27. $4.86 \times 10^{19}$ atoms  29. $9.46 \times 10^{12}$ km

## Chapter 13 FACTORING AND APPLICATIONS

### 13.1 Factors; The Greatest Common Factor

**Key Terms**

   1.   factoring       2.   factored form       3.   greatest common factor

   4.   factor

**Objective 1**

   1.   15           3.   3           5.   28

   7.   6

**Objective 2**

   9.   $2ab^2$       11.   $w^2 x^5 y^4$       13.   $9xy^2$

**Objective 3**

   15.   $-8a\left(a-3b-5c\right)$    17.   $10x\left(2x+4xy-7y^2\right)$    19.   $-13x^8\left(-2+x^4-4x^2\right)$

   21.   $8xy^2\left(7xy^2-3y+4\right)$

**Objective 4**

   23.   $\left(1+p\right)\left(1-q\right)$    25.   $\left(4x-y\right)\left(2x+3y\right)$    27.   $\left(4x-y^2\right)\left(3x^2-y\right)$

   29.   $\left(r^2+s^2\right)\left(3r-2s\right)$

### 13.2 Factoring Trinomials

### Key Terms

1. factoring     2. greatest common factor   3. prime polynomial

### Objective 1

1. prime       3. $(s-8)(s+4)$       5. $(x+9)(x+2)$

7. $(x-2)(x+1)$     9. $(x-7)(x+5)$     11. $(x-7y)(x-8y)$

13. $(q-6)(q+2)$     15. $(a-8b)(a-2b)$

### Objective 2

17. $2m(m-2)(m+1)$   19. prime       21. $3p^4(p+4)(p+2)$

23. $10k^4(k+5)(k+2)$            25. $x^3(x-2)(x-1)$

27. $2y^2(x-3y)(x+2y)$           29. $qr(r-7q)(r+3q)$

### 13.3 Factoring Trinomials by Grouping

### Key Terms

1. coefficient     2. trinomial

### Objective 1

1. $x + 3$

3. $4x - 2$

5. $(4b + 3)(2b + 3)$

7. $(5a + 2)(3a + 2)$

9. $(3b + 2)(b + 2)$

11. $p(3p + 2)(p + 2)$

13. $b(7a + 4)(a + 2)$

15. $(3c + 6d)(3c + 2d)$

17. $(5c - 7t)(2c - 3t)$

19. $(2x + 7y)(6x - 5y)$

21. $m(6m + 8n)(m - n)$

23. $(6f - g)(3f + 5g)$

25. $2a(4a - 5)(5a - 4)$

27. $4(2x + y)(x - y)$

29. $(2c + 3d)(5c + 12d)$

## 13.4    Factoring Trinomials Using FOIL

### Key Terms

1.  inner product    2.  FOIL    3.  outer product

### Objective 1

1.  $(2x+3)(5x+2)$    3.  $(2a+1)(a+6)$    5.  $(2m-3)(4m+1)$

7.  $(3q-4)(5q+6)$    9.  $(3w+2z)(3w+2z)$    11.  $(3x-4y)(2x+3y)$

13.  $(3y+5)(4y-3)$    15.  $(2p+1)(p+5)$    17.  $(9y+2)(y-2)$

19.  $(3r-1)(3r+5)$    21.  $2(2c-d)(c+4d)$    23.  $(9r-4t)(3r+2t)$

25.  $(7c-3d)(4c+5d)$    27.  $(3n-7s)(2n+9s)$

29.  $2a(3a+2b)(2a+3b)$

**13.5    Special Factoring Techniques**

**Key Terms**

   1.   difference of squares         2.   perfect square trinomial

**Objective 1**

   1.   $(5a - 6)(5a + 6)$     3.  $\left(3j - \frac{4}{7}\right)\left(3j + \frac{4}{7}\right)$     5.  $\left(4y^2 + 9\right)(2y - 3)(2y + 3)$

   7.   $m^2\left(mn - 1\right)\left(mn + 1\right)$

**Objective 2**

   9.   $(2x + 3)^2$         11.   $(4q - 5)^2$         13.   $\left(10p - \frac{5}{8}r\right)^2$

   15.   $\left[(p - q) - 10\right]^2$

**Objective 3**

   17.   $(x - y)(x^2 + xy + y^2)$     19.   $(6m - 5p^2)(36m^2 + 30mp^2 + 25p^4)$

   21.   $8(3x - y)(9x^2 + 3xy + y^2)$     23.   $3x^2 - 3x + 1$

**Objective 4**

   25.   $(3r + 2s)(9r^2 - 6rs + 4s^2)$     27.   $(5p + q)(25p^2 - 5pq + q^2)$

   29.   $2x(x^2 + 3y^2)$

### 13.6   A General Approach to Factoring
**Key Terms**

1.  factoring by grouping           2.   FOIL

**Objective 1**

1.   $-6x(2x+1)$           3.   $5rt(r-2+t)$           5.   $3a[a+2(x-y)]$

7.   $(m-n)^2$

**Objective 2**

9.   $2xy^2(xy+6)(xy-6)$

11.   $\left[(r+s)+2\right]\left[(r+s)^2-2(r+s)+4\right]$ or $\left[(r+s)+2\right]\left[r^2+2rs+s^2-2r-2s+4\right]$

13.   $\left(y^2+1\right)\left(y^4-y^2+1\right)$

15.   $\left[(3a-1)+y^3\right]\left[(3a-1)-y^3\right]$

**Objective 3**

17.   $(a-6)(2a-5)$       19.   $(5x-2y)(5x+y)$     21.   $(4m+5)(3m-1)$

23.   $(4b+1)(b+2)(b-1)$

**Objective 4**

25.   $(x-y)(a+b)$       27.   $(x-3)\left(x^2+7\right)$     29.   $(a-3b+5)(a-3b-5)$

**13.7   Solving Quadratic Equations by Factoring**

**Key Terms**

   1.   standard form     2.   quadratic equation

**Objective 1**

   1.   $\{-2, -5\}$      3.   $\{-7, 7\}$      5.   $\{-7, 9\}$

   7.   $\left\{-\frac{2}{3}, 3\right\}$      9.   $\left\{-\frac{2}{3}, -\frac{2}{3}\right\}$      11.   $\left\{-\frac{4}{3}, \frac{4}{3}\right\}$

   13.   $\left\{-\frac{2}{7}, \frac{3}{2}\right\}$      15.   $\left\{-3, \frac{1}{2}\right\}$

**Objective 2**

   17.   $\left\{-\frac{3}{2}, 0, 5\right\}$      19.   $\{-7, 0, 7\}$      21.   $\{-4, 0, 2\}$

   23.   $\left\{-\frac{3}{2}, \frac{3}{2}, 2\right\}$      25.   $\left\{-\frac{4}{3}, 0, 1\right\}$      27.   $\{-6, 2, 3, 6\}$

## 13.8   Applications of Quadratic Equations

### Key Terms

1.   legs          2.   hypotenuse

### Objective 1

1.   width: 8 in., length: 24 in.

3.   rectangle 1: width: 4 m, length: 12 m; rectangle 2: width: 6 m; length: 8 m

5.   base: 12 cm; height: 7 cm

7.   width: 3m, length: 5 m

### Objective 2

9.   −4, −3 or 3, 4          11.   12, 14          13.   8, 10

15.   6 in., 8 in., 10 in.

### Objective 3

17.   car: 60 mi; train: 80 mi

19.   45 mi          21.   20 mi

### Objective 4

23.   a. $\frac{1}{2}$ sec; b. 1 sec; c. 2 sec

25.   110 items          27.   40 items or 110 items

29.   404 ft

# Chapter 14 FACTORING AND APPLICATIONS

## 14.1 The Fundamental Property of Rational Expressions

### Key Terms

1. rational expression   2. lowest terms

### Objective 1

1. $x \neq -7$          3. $x \neq -5, x \neq 5$          5. $z \neq 3$

7. $y \neq -2, y \neq 2$

### Objective 2

9. a. $-\frac{11}{9}$; b. $-2$     11. a. $\frac{3}{25}$; b. $-\frac{9}{121}$     13. a. $-\frac{3}{10}$; b. $\frac{11}{10}$

### Objective 3

15. $\dfrac{-5b}{8c}$          17. $4k$          19. $\dfrac{9(x+3)}{2}$

21. $\dfrac{6r+5s}{r+5s}$     23. $\dfrac{v+3}{v-2}$

### Objective 4

25. $\dfrac{-(4x+5)}{3-6x}; \dfrac{-4x-5}{3-6x}; \dfrac{4x+5}{-(3-6x)}; \dfrac{4x+5}{-3+6x}$

27. $\dfrac{-(2x-3)}{x+2}; \dfrac{-2x+3}{x+2}; \dfrac{2x-3}{-(x+2)}; \dfrac{2x-3}{-x-2}$

29. $\dfrac{-(2x-1)}{3x+5}; \dfrac{-2x+1}{3x+5}; \dfrac{2x-1}{-(3x+5)}; \dfrac{2x-1}{-3x-5}$

### 14.2 Multiplying and Dividing Rational Expressions

### Key Terms

1. reciprocal     2. rational expression     3. lowest terms

### Objective 1

1. $\dfrac{10m^3n}{3}$     3. $-\dfrac{2}{3}$     5. $-\dfrac{1}{8+2x}$

7. $-\dfrac{x-2}{4x+4}$     9. $\dfrac{x-3}{x+2}$     11. $\dfrac{x+5}{x-9}$

### Objective 2

13. $\dfrac{2y}{7}$     15. $\dfrac{1}{(2r+1)(2r+3)}$     17. $\dfrac{2x^2+9}{(x-3)(x+1)}$

### Objective 3

19. $-\dfrac{1}{2}$     21. $\dfrac{2(m-5)}{m+3}$     23. $\dfrac{4a+3}{a-4}$

25. $\dfrac{(y-3z)(y-6z)}{(y-4z)(y+5z)}$     27. $\dfrac{4(y-1)}{3(y-3)}$     29. $\dfrac{2k-3}{k-1}$

## 14.3    Least Common Denominators

### Key Terms

1.  equivalent expressions

2.  least common denominator

### Objective 1

1.  $48r^4$        3.  $108b^4$        5.  $42(t-4)$

7.  $a^2-b^2$ or $b^2-a^2$  9.  $a(a-2)(2a+5)$    11.  $(t+2)(t+4)(t-3)$

13.  $(2q-5)(q+2)(q-2)$        15.  $m^2(m-2)(m+7)$

### Objective 2

17.  $24w$            19.  8            21.  $44a+4$ or $4(11a+1)$

23.  $8z(4z+1)$ or $32z^2+8z$        25.  $9(y+2)=9y+18$

27.  $2(3x+1)$ or $6x+2$          29.  $3(k+7)$ or $3k+21$

### 14.4 Adding and Subtracting Rational Expressions

**Key Terms**

1. greatest common factor

2. least common denominator

**Objective 1**

1. $\dfrac{4}{w^2}$

3. $\dfrac{1}{b-2}$

5. $\dfrac{1}{2y+1}$

**Objective 2**

7. $\dfrac{11x+15}{(x-5)(x+5)}$

9. $\dfrac{5m^2+12m+10}{(m-4)(m+4)(m+1)}$

11. $\dfrac{16s^2+19s-1}{(3s-2)(s-4)(2s+3)}$

13. $\dfrac{7z^2-z-6}{(z+2)(z-2)^2}$

15. $\dfrac{x+7}{(3x+2)(2x-1)(x+2)}$

17. $\dfrac{c^2+3c-16}{(2c+3)(c-1)(c-4)}$

**Objective 3**

19. $\dfrac{8z}{(z-2)(z+2)}$ or $\dfrac{8z}{z^2-4}$

21. $\dfrac{3x-2}{2(x-2)(x+2)}$

23. $-\dfrac{1}{2d+3}$

25. $\dfrac{2m^2-m+2}{(m-2)(m+2)^2}$

27. $\dfrac{2c-16}{(2c+3)(c+2)(c-2)}$ or $\dfrac{2(c-8)}{(2c+3)(c+2)(c-2)}$

29. $\dfrac{11x^2-x-11}{(2x-1)(x+3)(3x+2)}$

**14.5 Complex Fractions**

**Key Terms**

1. complex fraction   2. LCD

**Objective 1**

1. $-\dfrac{2}{3}$

3. $\dfrac{7m^2}{2n^3}$

5. $\dfrac{2y-5}{3y-8}$

7. $\dfrac{3p-2}{2p+1}$

9. $\dfrac{9s+12}{6s^2+2s}$ or $\dfrac{3(3s+4)}{2s(3s+1)}$

11. $(a+2)^2$

13. $\dfrac{24}{w-3}$

15. $\dfrac{5(3a+4)}{2a+5}$

**Objective 2**

17. $-\dfrac{7}{3}$

19. $\dfrac{4}{r^2s^2}$

21. $\dfrac{r^2+3}{5+r^2t}$

23. $\dfrac{2s^2+3}{1-3s^2}$

25. $\dfrac{2(1-4h)}{h(1+4h)}$

27. $\dfrac{4m-3}{2(3-2m)}$

29. $\dfrac{(k-23)(k+2)}{5k(k+1)}$

## 14.6    Solving Equations with Rational Expressions

### Key Terms

1.    expression    2.    equation

### Objective 1

1.    equation; {4}    3.    operation; $\dfrac{3x}{10}$    5.    operation; $\dfrac{41x}{15}$

### Objective 2

7.    $\{-11\}$    9.    $\{-4, 16\}$    11.    $\left\{\dfrac{41}{5}\right\}$

13.    $\left\{\dfrac{1}{17}, 7\right\}$    15.    $\{-2, 1\}$    17.    $\{2\}$

19.    $\{-24, 1\}$    21.    $\{-1, 3\}$

### Objective 3

23.    $\dfrac{s - a_1}{s}$    25.    $y_2 - m\left(x_2 - x_1\right)$    27.    $\dfrac{Fd^2}{m_1 m_2}$

29.    $\dfrac{2A}{h} - b_1$ or $\dfrac{2A - b_1 h}{h}$

### 14.7 Applications of Rational Expressions

**Key Terms**

1. numerator
2. denominator
3. reciprocal

**Objective 1**

1. $-\frac{2}{3}$ or 1
3. $\frac{9}{13}$
5. $\frac{1}{3}$ or $\frac{3}{4}$
7. $\frac{3}{5}$
9. 2, 4

**Objective 2**

11. 5 miles per hour
13. 150 miles per hour
15. 2 miles per hour
17. 50 miles per hour
19. 380 miles per hour

**Objective 3**

21. $2\frac{2}{5}$ hr
23. $\frac{2}{5}$ hr
25. 3 hrr
27. 6 hr
29. 72 hr

**14.8   Variation**

**Key Terms**

1.   constant of variation                    2.   direct variation

3.   inverse variation

**Objective 1**

1.   72                    3.   $\frac{68}{3}$ or $22\frac{2}{3}$                    5.   15

7.   9                    9.   168

**Objective 2**

11.   $\frac{20}{3}$ or $6\frac{1}{3}$                    13.   4                    15.   2.25

17.   6                    19.   3                    21.   $162.50

23.   200 ft$^3$                    25.   69.08 cm                    27.   12.8 cm

29.   275 mi

# Chapter 15 SYSTEMS OF LINEAR EQUATIONS AND INEQUALITIES

## 15.1 Solving Systems of Linear Equations by Graphing

### Key Terms

1. independent equations

2. consistent system

3. solution set of the system

4. solution of a system

5. dependent equations

6. inconsistent system

7. system of linear equations

### Objective 1

1. no

3. yes

5. yes

7. yes

### Objective 2

9.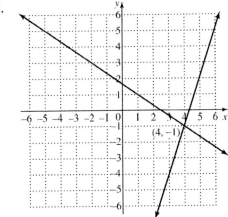

11.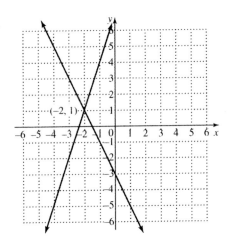

13.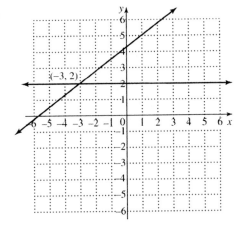

15.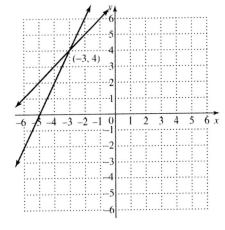

## Objective 3

17.   infinite number of solutions

19.   infinite number of solutions

21.   no solution

## Objective 4

23.   (a) neither      (b) intersecting lines      (c) one solution

25.   (a) dependent      (b) one line      (c) infinitely many solutions

27.   (a) neither      (b) intersecting lines      (c) one solution

29.   (a) neither      (b) intersecting lines      (c) one solution

**15.2   Solving Systems of Linear Equations by Substitution**

**Key Terms**

1.   ordered pair          2.   substitution          3.   dependent system

4.   inconsistent system

**Objective 1**

1.   (1, 6)          3.   (2, 7)          5.   (−2, −1)

7.   (−1, −7)          9.   $\left(3, -\dfrac{2}{3}\right)$          11.   (−2, −3)

**Objective 2**

13.   ∅          15.   ∅          17.   ∅

**Objective 3**

19.   (6, 2)          21.   (2, 3)          23.   (6, 2)

25.   (7, −2)          27.   (4, −9)

29.   $\{(x, y) \mid 0.3x + 0.4y = 0.5\}$

## 15.3 Solving Systems of Linear Equations by Elimination

### Key Terms

1. elimination method  2. addition property of equality

3. substitution

### Objective 1

1. $(1, 4)$          3. $(8, 3)$          5. $(2, -4)$

7. $(5, 0)$

### Objective 2

9. $\left(\dfrac{1}{2}, 1\right)$      11. $(-4, 4)$      13. $(-3, 2)$

15. $\left(\dfrac{1}{2}, -\dfrac{3}{2}\right)$

### Objective 3

17. $(-3, -2)$      19. $(-4, 1)$      21. $(3, -2)$

23. $(-2, 2)$

### Objective 4

25. $\varnothing$          27. $\{(x, y) \mid 2x - 4y = 1\}$

29. $\{(x, y) \mid 48x - 56y = 32\}$ or $\{(x, y) \mid -18x + 21y = -12\}$

## 15.4 Applications of Linear Systems

### Key Terms

1. $d = rt$      2. system of linear equations

### Objective 1

1. 12, 8      3. 5647 people, 3398 people

5. 56 cm, 26 cm      7. 32 cm, 32 cm, 52 cm

### Objective 2

9. 30 \$5 bills; 60 \$10 bills

11. \$6000 at 7%; \$4000 at 4%

13. \$4000 at 7%; \$8000 at 9%

15. 8 \$14 ties; 2 \$25 ties

### Objective 3

17. water: 6 liters; 25% solution: 24 liters

19. \$90 coffee: 40 bags; \$75 coffee: 10 bags

21. water: 9 oz; 80% solution: 3 oz

### Objective 4

23. Bill: 642 kph; Hillary: 582 kph

25. Enid: 44 mph; Jerry: 16 mph

27. John: 54 mph; Mike: 52 mph

29. plane speed: 265 mph; wind speed: 35 mph

### 15.5 Solving Systems of Linear Inequalities

### Key Terms

1. solution set of a system of linear inequalities

2. system of linear inequalities

### Objective 1

1.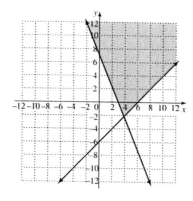

3.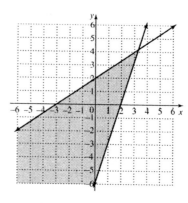

5.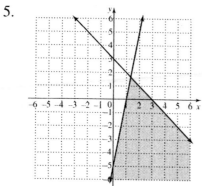

7.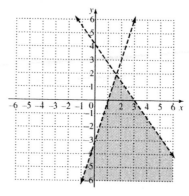

9.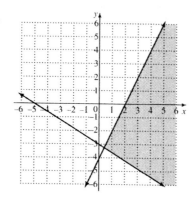

11.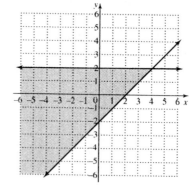

13.

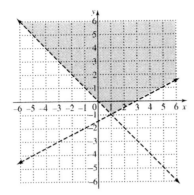

15.

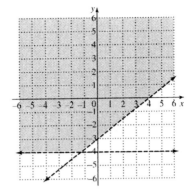

17.

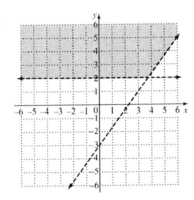

19.

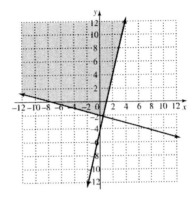

21.

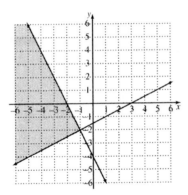

23.

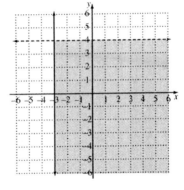

25.

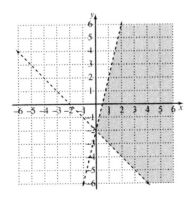

27.

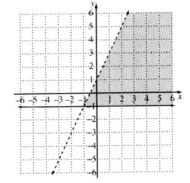

29.

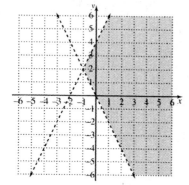

# Chapter 16 ROOTS AND RADICALS

## 16.1    Evaluating Roots

### Key Terms

| | | | | | |
|---|---|---|---|---|---|
| 1. | radicand | 2. | perfect square | 3. | index (order) |
| 4. | square root | 5. | radical expression | 6. | principal square root |
| 7. | irrational number | 8. | radical | 9. | cube root |

### Objective 1

1.    25, −25            3.    $\frac{11}{14}, -\frac{11}{14}$            5.    $\frac{30}{7}$

### Objective 2

7.    irrational            9.    rational            11.    irrational

### Objective 3

13.    5.657            15.    14.142            17.    14.491

### Objective 4

19.    10.296            21.    5.385            23.    24 feet

### Objective 5

25.    −4            27.    −3            29.    4

## 16.2    Multiplying, Dividing, and Simplifying Radicals
### Key Terms

1.  radical                    2.  perfect cube                    3.  radicand

### Objective 1

1.  $\sqrt{65}$                 3.  $\sqrt{300}$                    5.  $\sqrt{21x}$

### Objective 2

7.  $-6\sqrt{5}$               9.  $30\sqrt{10}$                  11.  $10\sqrt{2}$

### Objective 3

13.  $\dfrac{5}{9}$            15.  1                             17.  $\dfrac{2}{25}$

### Objective 4

19.  $pq^3$                    21.  $4x^2 y^2 \sqrt{2y}$          23.  $\dfrac{9}{5x^3}$

### Objective 5

25.  $-2\sqrt[5]{2}$          27.  $2\sqrt[4]{4}$               29.  $\dfrac{5}{4}$

## 16.3   Adding and Subtracting Radicals

### Key Terms

1. index

2. unlike radicals

3. like radicals

### Objective 1

1. $6\sqrt{11}$

3. $\sqrt{2}$

5. $8\sqrt{3}$

7. $6\sqrt{2}$

9. $4\sqrt[5]{4}$

### Objective 2

11. $40\sqrt{2}$

13. $-6\sqrt[3]{2}$

15. $-4\sqrt{2}+4\sqrt{5}$

17. $15\sqrt{2}$

19. $\sqrt[4]{2}$

### Objective 3

21. $4\sqrt{35}$

23. $17\sqrt{5x}$

25. $39w\sqrt{6}$

27. $-\sqrt{3y}$

29. $-4\sqrt{66}$

## 16.4    Rationalizing the Denominator

### Key Terms

1. product rule    2. rationalizing the denominator

3. quotient rule

### Objective 1

1. $-\sqrt{3}$

3. $-\dfrac{2\sqrt{3}}{3}$

5. $\dfrac{\sqrt{2}}{4}$

7. $\dfrac{\sqrt{6}}{6}$

9. $\dfrac{5\sqrt{2}}{2}$

### Objective 2

11. $\dfrac{\sqrt{15}}{3}$

13. $\dfrac{3\sqrt{14}}{10}$

15. $\dfrac{x^3}{y^2}$

17. $\dfrac{ab\sqrt{30b}}{6}$

19. $\dfrac{2\sqrt{6qt}}{t}$

### Objective 3

21. $\dfrac{\sqrt[3]{28}}{2}$

23. $\dfrac{3\sqrt[3]{7}}{7}$

25. $\dfrac{\sqrt[3]{36}}{18}$

27. $\dfrac{\sqrt[3]{5x^2y^2}}{5y}$

29. $\dfrac{\sqrt[3]{35x^2}}{7x}$

### 16.5   More Simplifying and Operations with Radicals

**Key Terms**

1.  rationalize the denominator

2.  conjugate

**Objective 1**

1.  $2\sqrt{15} + 4\sqrt{35}$

3.  $4\sqrt{10} - 4\sqrt{35} + \sqrt{6} - \sqrt{21}$

5.  $24 + 2\sqrt{15} - 8\sqrt{30} - 10\sqrt{2}$

7.  $-38$

9.  $97 + 56\sqrt{3}$

**Objective 2**

11.  $-4\sqrt{3} + 8$

13.  $-\sqrt{5} - \sqrt{2}$

15.  $\dfrac{-5\left(\sqrt{3} + \sqrt{10}\right)}{7}$

17.  $\dfrac{\sqrt{3} + 2\sqrt{6} + \sqrt{2} + 4}{-7}$

19.  $-\sqrt{15} + 2\sqrt{5} + 2\sqrt{3} - 4$

**Objective 3**

21.  $\dfrac{\sqrt{7} - 2\sqrt{2}}{3}$

23.  $\dfrac{2\sqrt{2} - 3}{3}$ or $\dfrac{2\sqrt{2}}{3} - 1$

25.  $\dfrac{1 + \sqrt{3}}{3}$

27.  $\sqrt{15} + \sqrt{10} + 4\sqrt{3} + 4\sqrt{2}$

29.  $27\sqrt{3} + 5$

## 16.6 Solving Equations with Radicals

### Key Terms

1. extraneous solution  2. radical equation

### Objective 1

1. $\left\{\dfrac{8}{3}\right\}$

3. $\left\{\dfrac{2}{5}\right\}$

5. $\left\{\dfrac{1}{3}\right\}$

7. $\{2\}$

9. $\{-3\}$

### Objective 2

11. $\varnothing$

13. $\varnothing$

15. $\left\{-\dfrac{4}{17}\right\}$

17. $\varnothing$

### Objective 3

19. $\{8\}$

21. $\{3\}$

23. $\{-1\}$

25. $\{5\}$

27. $\{4\}$

### Objective 4

29. 499 sq cm

## Chapter 17 QUADRATIC EQUATIONS

### 17.1 Solving Quadratic Equations by the Square Root Property

### Key Terms

1. quadratic equation   2. zero-factor property

### Objective 1

1. $\{-30, 30\}$

3. $\varnothing$

5. $\{-1.4, 1.4\}$

7. $\{-5.5, 5.5\}$

9. $\left\{-\dfrac{3\sqrt{10}}{17}, \dfrac{3\sqrt{10}}{17}\right\}$

### Objective 2

11. $\{-6, 2\}$

13. $\left\{-\dfrac{13}{7}, 3\right\}$

15. $\left\{4 - \sqrt{7},\ 4 + \sqrt{7}\right\}$

17. $\left\{\dfrac{1}{5}, \dfrac{4}{5}\right\}$

19. $\{-26, 10\}$

21. $\varnothing$

### Objective 3

23. $\dfrac{5\sqrt{15}}{2}$ or about 9.7 ft

25. 2 in.

27. width: 10 ft; length 20 ft

29. $5\sqrt{3}$ or about 8.7 cm

**17.2    Solving Quadratic Equations by Completing the Square**

**Key Terms**

1.  perfect square trinomial

2.  square root property

3.  completing the square

**Objective 1**

1.  $\{-4, 1\}$

3.  $\left\{-2 - \sqrt{6},\ -2 + \sqrt{6}\right\}$

5.  $\{-9, 7\}$

7.  $\left\{\dfrac{1 - \sqrt{11}}{2},\ \dfrac{1 + \sqrt{11}}{2}\right\}$

**Objective 2**

9.  $\left\{\dfrac{3 - \sqrt{15}}{3},\ \dfrac{3 + \sqrt{15}}{3}\right\}$

11.  $\left\{-\dfrac{3}{2}, \dfrac{5}{3}\right\}$

13.  $\{-4, 2\}$

15.  $\left\{\dfrac{-2 - \sqrt{10}}{6},\ \dfrac{-2 + \sqrt{10}}{6}\right\}$

**Objective 3**

17.  $\left\{-\dfrac{3}{2}, \dfrac{1}{2}\right\}$

19.  $\left\{\dfrac{1 - \sqrt{3}}{2},\ \dfrac{1 + \sqrt{3}}{2}\right\}$

21.  $\left\{-2 - \sqrt{2},\ -2 + \sqrt{2}\right\}$

**Objective 4**

23.  3 sec

25.  0.05 or 5%

27.  1 sec and 4 sec

29.  2 months, 4 months

## 17.3 Solving Quadratic Equations by the Quadratic Formula

### Key Terms

1. standard form    2. constant    3. quadratic formula

### Objective 1

1. $a = 10$, $b = 4$, $c = 0$    3. $a = 3$, $b = 0$, $c = -12$    5. $a = 1$, $b = 1$, $c = -9$

### Objective 2

7. $\{-13, 1\}$    9. $\varnothing$    11. $\left\{ \dfrac{-2 - \sqrt{14}}{5}, \dfrac{-2 + \sqrt{14}}{5} \right\}$

13. $\left\{ \dfrac{7}{6} \right\}$

### Objective 3

15. $\{2\}$    17. $\left\{ \dfrac{1}{3} \right\}$    19. $\{9\}$

21. $\left\{ -\dfrac{3}{10} \right\}$

### Objective 4

23. $\left\{ -1 - \sqrt{17}, \ -1 + \sqrt{17} \right\}$    25. $\varnothing$

27. $\{-3, 1\}$    29. $\dfrac{2 - \sqrt{14}}{2}, \dfrac{2 + \sqrt{14}}{2}$

## 17.4    Graphing Quadratic Equations

### Key Terms

1.   line of symmetry    2.   vertex        3.   axis

4.   parabola

### Objective 1; Objective 2

1.

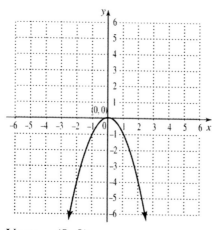

Vertex: (0, 0)

3.

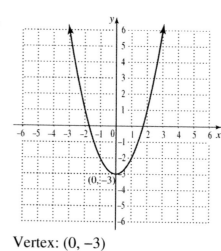

Vertex: (0, −3)

5.

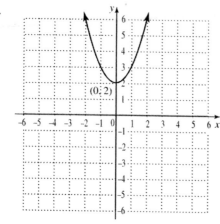

Vertex: (0, 2)

7.

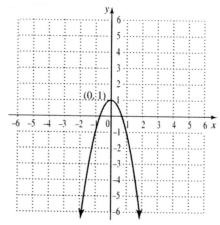

Vertex: (0, 1)

9.

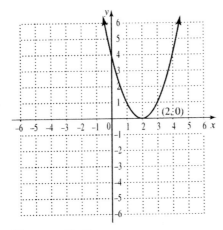

Vertex: (2, 0)

11.

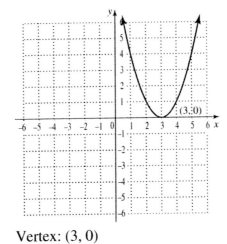

Vertex: (3, 0)

13.

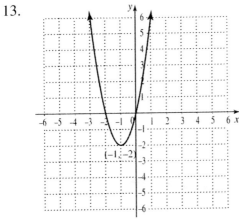

Vertex: (−1, −2)

15.

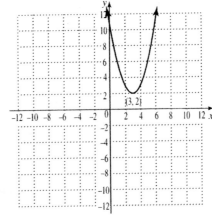

Vertex: (3, 2)

17.

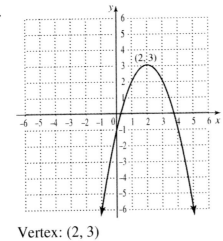

Vertex: (2, 3)

19.

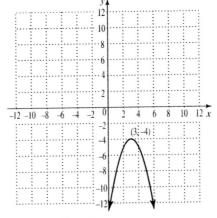

Vertex: (3, −4)

21.

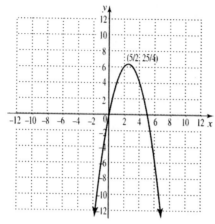

Vertex: $\left(\dfrac{5}{2}, \dfrac{25}{4}\right)$

23.

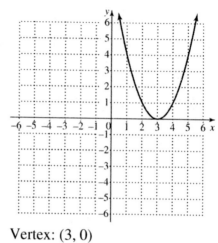

Vertex: $(3, 0)$

25.

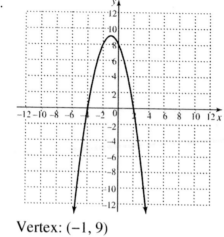

Vertex: $(-1, 9)$

27.

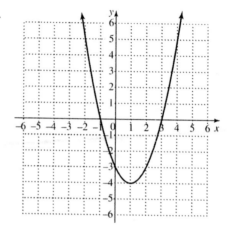

Vertex: $(1, -4)$

29.

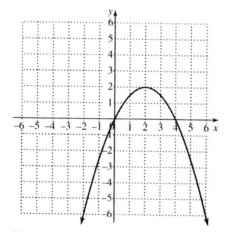

Vertex: $(2, 2)$

## 17.5   Introduction to Functions

### Key Terms

1. relation
2. range
3. function
4. components
5. domain

### Objective 1; Objective 2

1. function; domain: $\{-3, 0, 2, 5\}$; range: $\{-8, -4, -1, 2, 7\}$

3. not a function; domain: $\{-7, -3, 0, 2\}$; range: $\{1, 4, 6, 7, 9\}$

5. function; domain: $\{-5, -2, 1, 3, 7\}$; range: $\{4\}$

7. function; domain: $\{-3, -2, -1, 0, 1\}$; range: $\{-5, 0, 5\}$

9. not a function; domain: $\{A, B, C, D, E\}$; range: $\{50, 60, 70, 80, 90\}$

11. function

### Objective 3

13. function
15. not a function
17. function

### Objective 4

19. (a) $-13$; (b) $-7$; (c) $5$
21. (a) $1$; (b) $-5$; (c) $31$

23. (a) $-12$; (b) $4$; (c) $36$

### Objective 5

25. $\{(2003, 719 \text{ million}), (2004, 817 \text{ million}), (2005, 1018 \text{ million}), (2006, 1093 \text{ million}), (2007, 1262 \text{ million})\}$; yes

27. $w(2003) = 719$ million; $w(2006) = 1093$ million

29. $\{(2001, 596 \text{ thousand}), (2002, 625 \text{ thousand}), (2003, 872 \text{ thousand}), (2004, 795 \text{ thousand}), (2005, 625 \text{ thousand})\}$; function